MINIMALIST
PACKAGING

• • • •

Enhancing Creative Concepts

MINIMALIST

PACKAGING

ENHANCING CREATIVE CONCEPTS

images
Publishing

So Simple!

C O N T E N T S

Case Studies

So Simple!

By Chris Huang

Chris Huang is an award-winning marketing design professional, the owner of a graphic design agency, and an art college professor in Chicago. She specializes in sales and marketing communication design solutions. Her client list includes: McDonald's Corporation, Hyatt Hotel, FedEx, Harris Bank, HSBC Household, Abbott Laboratory, and others. Chris Huang is also the author of *Fashion Packaging Now*.

1 What is minimalism?

Minimalism has become one of the most important and influential styles since World War I. With the decline and destruction of most major empires, traditional and complex styles were no longer valued;[1] this aspect of society had changed. New and contemporary art theories, coupled with revolutions in typography and new terms in "graphic design," were steadily introduced. Extending from the concept of "forms follows functions" touted by Bauhaus and visually inspired by the geometric and abstract, minimalism became a favorite belief of architects, musician, designers, and artists.

"Less is more." This is the most popular and representative aphorism from the former director of Bauhaus Schools, German–American architect Ludwig Mies van der Rohe (1886–1969), clearly illuminating the essential theory of simplicity, of minimalism. Minimalism is not all about simplicity. However, simplicity is the heart of minimalism. In the last 50 years of the minimalism movement, the definition of simplicity has shifted to a different time and space and now widely influences all design-related industries, including packaging.

In 2004, Japanese industrial designer Naoto Fukasawa designed the JUICEPEEL packaging campaign, which simply used fruit skins with texture to successfully illustrate the product inside. This brilliant and minimalist packaging design played the primary role of communicator to anyone, both those who could understand Japanese and those who couldn't.

2 Advocate for minimalist packaging design: Apple

For a decade, Apple has been ranked No. 1 on Forbes' list of the World's Most Valuable Brands.[2] The secret of Apple's success is simplicity,[3] driven by the vision of its CEO, Steve Jobs. Inspired by Frank Lloyd Wright's vision of simple, modern homes and deeply influenced by Japanese Zen Buddhism's minimalist aesthetics, Jobs repeatedly emphasized that Apple's mantra would be simplicity, which was echoed by Apple's Design Chief[4] Johnathan Ive who said, "My work is all about simplicity."

Apple's iPhone is not a consumer brand, it's a fashion icon.[5] Relevant design industries swarmed to the style of simplicity just to be like Apple. To remain consistent with Apple's product design concept and to tell the brand's complete story, Apple designed its packaging artistically to be as visually appealing as the product inside. The simplicity of the iPhone's packaging became a major draw for consumers as well. Minimalist packaging design reached its peak because of Apple.

3 Why does minimalist packaging design work?

Minimalist theory has been around for over half a century, and Apple has raised the bar in packaging design for over a decade. The trend of minimalism in packaging design continues to pick up steam. Why do packaging designers strive for simplistic designs? What are the benefits of having minimalist packaging designs?

Eye-catching

Information easily overwhelms the consumers in today's increasingly connected world. According to Statista.com, there was a 92 percent growth in supermarket sales in the United States compared to 15 years ago.[6] In Western Europe, 12,000 innovations were launched in four markets across 17 product categories between 2011 and 2013.[7] Today's consumers have many more brands and products to choose from. Minimalist packaging design helps consumers to focus on what's important. Without superfluous design elements that distract consumers from the central focus, products with minimalist packaging designs tend to stand out from the competitors.

On trend

Research has shown that the arts influence society by changing opinion, instilling values and translating experience across time and space. The arts are also an accurate showcase of social movements,[8] whether it moves from society to art or art to society; they are inseparable. When a new art form theory is born, societal behavior and all fine arts are influenced by each other. The mass market is seeing this now, with the influence of the internet on how consumers shop.

With the advent of e-commerce, brand competition is getting intense. Vying for attention on web pages and store shelves are almost alike. User behavior from surfing the web and using cell phones changes everything. Jakob Nielsen's eye-tracking study showed that less than 20 percent of the text content is actually read on an average web page.[9] People don't read—they scan. Less is better. Minimal web page design is more necessary than ever, and packaging is the same.

Minimalist packaging is the trend. "Simple, clean, minimal" was illustrated in the predictions of product packaging design trends by many professional packaging design bloggers:

"Minimalistic packaging in 2019 will focus on clean and simple designs that let color and typography take center stage—which is incredibly impactful and guaranteed to stand out."

—Martic Lupus: 9 Inspiring Packaging Design Trend for 2019[10]

"Packaging Design Trends: Minimalism: Less is more! Though it may be tempting to feature a lot of text or different fonts in your packaging, this can create a cluttered design."

—Loredana Papp-Dinea and Mihai Baldean: Behance: Trend of the year 2019[11]

"Minimalist packaging design is not necessarily a new trend, but over time, we've seen it pick up more and more momentum. In 2019, we'll watch it continue to gain popularity."

—David Roberge, A Predictive Look Ahead: 2019 Product Packaging Trends[12]

"Keeping things minimal has been the order of the day for a very long time, and feedback from buyers reveals that this is not going anywhere, at least for now."

—Inkbot Design, 10 Creative Packaging Design Trends for 2019[13]

"2018 was a fantastic year of minimalism and functionality. We foresee the same thing happening in 2019."

—Source Nutraceutical, Inc., New Year, New Packaging Design Trend[14]

Ultimate sophisticate

Influenced by Baroque and Rococo styles, rich fashion was defined by extreme refinement in the 18th century. Nowadays, the definition of sophistication has shifted completely from Baroque's complexity to minimalism's simplicity. Minimalist sensibilities not only highlight the product as a focal point, but also feature product quality cleverly. It pushes the brand to a higher, superior position.

"Simplicity is the ultimate sophistication," said Leonardo Da Vinci, a sentiment that was later reemphasized by Steve Jobs. Apple's iPhone is the number-one way to guess if someone is rich. A National Bureau of Economic Research paper produced by economists from the University of Chicago found that 69.1 percent of respondents with an iPhone belonged to the top quartile of households by income.

Apple's sophisticated, clean, elegant, black-and-white packaging symbolizes luxury. To emphasize fine jewelry's beauty, advertisers often use a simple black-and-white picture as a background for contrast. Some luxury brands' packaging even feature their brands without any words and only show their brand logos with nothing else. Simplicity is sophisticated, elegant, and luxurious.

Sustainability strategy

Enhanced by the concept of eliminating needless materials with the 3Rs—reduce, reuse, recycle—environmentally friendly purposes empower packaging design to push to the top of the minimalism pyramid. In the hierarchy of the 3Rs, reduce comes first; it's best to prevent waste in the first place. Less material reduces cost, fitting minimalism's mantra of "less is more."

According to Luminer's 2017 consumer survey statistic, 56 percent of shoppers are more likely to choose a product that uses eco-friendly/sustainable packaging over a similar product that does not.[16] In 'The Power of Packaging: 2018 State of Industry Report' by BRANDPackaging, 57 percent of packaging professionals believe that sustainability and recyclability are the top packaging trends. These professionals also look at using less BPA, Styrofoam and plastics materials.[17]

Minimalist packaging design with an emphasis on recyclable and biodegradable materials and optimization of material usage not only sets you apart from your competition but will also appeal to the eco-conscious consumers as well. It's also a great brand image for public relations.

Cost effective

The strategy of overpackaging to impress shoppers is outdated. The fancy materials, extra printing and labor process or unnecessary layers all only end up in landfills.

Besides the fundamental materials needed for drop protection, how many other needless things can be eliminated? Using a large box for a small item is not only a waste of space and material but also affects the entire distribution process from warehouse storage to retail display.

Today's smarter consumers are aware that they are paying for what they are given. Using less materials leads to lower overheads due to less labor and lower shipping costs. According to a new survey by NPR (National Public Radio) and the Marist Institute for Public Opinion in 2018, nearly 7 in 10 Americans (69 percent) say they have purchased an item online.[18] The larger and heavier the object, the higher the shipping cost. Minimizing packaging layers and size and choosing lightweight materials helps to reduce shipping cost, a huge plus for competitive e-commerce businesses. Less is truly better. Less packaging is cost effective.

Silent salesman

Product packaging is a silent salesman. The best salesmen can quickly and easily communicate an idea that facilitates the consumer's buying decision. When placed next to competitors, packaging conveys product advantages and allows potential buyers to gain the information that may differentiate it from others.

Having a pithy and eye-grabbing presentation helps sales. Proper packaging design is the same. In Luminer's August 2017 survey of 400 American men and women (average

age of 46), 33 percent of shoppers are inclined to reject a product when they don't like the label. The power of a label influences shoppers' buying decisions. A brilliant, minimalist design packaging acts as a sales force sitting on the shelves among all competitors, waving to shoppers and saying, "Pick me!" Packaging's function should not only be for protection but selling as well.

Minimalist packaging design is classy and easily grabs consumers' attention to promote sales. In addition to using less and eco-friendly materials, minimalist design is cost savings and on the trend. Nowadays, simplistic style is widely applied when redesigning packaging for being competitive and up-to-date. Is minimalism right for your next packaging design?

4 How to design minimalist packaging

How do you design a minimalist package? Keep this principle in mind:

Omit needless things and focus on what is really important.

Identifying what is important and what is needless is the first step and the most critical process. Less is not really less. Simplicity is not really simple. Strategically creating every design element and word to serve an essential purpose tends to result in clean and direct visuals.

The following outlines minimalist design principles in each visual category.

Less elements

What are the must-have contents? Different industries have different requirements and standards. Using food products as an example, to help protect public health and safety, food labeling requirements vary in different countries by law. What is commonly required?

- **Product name**—The statement of identity is the name of the food. It must appear on the front label.

- **Net weight or volume**—The net quantity of contents (net quantity statement) is the statement on the label that provides the amount of product in the container or package. It must be expressed in weight, measure, or numeric count.

- **Date mark**—This regulation makes it mandatory that the information about the 'Date of manufacture or packing' and 'Best before' or 'Use by date' must be mentioned on the label.

- **Ingredient list**—The ingredients are listed in order of predominance, with the ingredients used in the greatest amount first, and followed in descending order by those in smaller amounts. The label must list the names of any FDA-certified color additive.

- **Nutrition facts information**—The nutrition facts label provides consumers with quick food choice information containing serving size, calories, nutrients, and percentage daily value.

- **Warning or declaration**—If an advisory statement is required due to safety of using certain products like drug and chemical products, then designers should add one. "Warning" and "Declaration" should appear on the label prominently and conspicuously as compared to other words.

- **Manufacturer information**—Declaration of responsibility. Name and address of manufacturer, distributor, or importer.

- **Country of origin**—In the United States, the USDA requires mandatory Country of Origin Labeling (COOL). All imported products must be marked with where products were manufactured, produced, and grown; and it's required to clear customs.

- **Storage instructions**—If a product requires specific storage instructions to remain safe, it's important to list those on the packaging label, such as "store in a cool dry place" or "keep in the fridge once opened."

- **Lot or batch number**—Lot or batch number is printed on the labels clearly at the end of the manufacturing process. It enables tracing of the product's entire manufacturing history of constituent parts or ingredients as well as labor and equipment record. It's a very important information when it comes to product recalls.

Requirements of packaging labeling in other industries and countries may be different. Besides the requirement above, there are a few other advantages that are not required that just can't be missed, such as:

- **Brand logo**—Brand names influence purchase decisions. Nielsen's 'Global New Product Innovation Survey' found that more than two-thirds of developing-market respondents (68 percent) say they prefer to buy new products from brands they're familiar with.[19] A 2017 survey from Luminer showed that 56 percent of shoppers say a recognizable logo draws their attention to a product.[20] That's why the brand name is at the top of the design hierarchy.

- **Certification icons**—USDA Organic, Fair trade, Kosher, Non GMO, FDA approved, etc. Getting certified is important. Product certification indicates the product's suitability and safety and that it has passed performance tests and quality assurance tests. Having an authorized third-party's guarantees is word-of-mouth marketing.

- **Barcode**—A barcode may not be required by law, but it was commonly requested by distribution for inventory controls. A barcode is a unique identifier on retail products that helps to track sales of products.

Other content may be promotional marketing messages. Other superfluous designs to fill in the blanks for balance may help but may also be eliminated to adhere more closely to minimalist design principles.

- **Advertising tagline**—A successful catchy tagline grabs the target audience's attention and influences the customer's buying decision. "New!", "Sugar Free," "No Trans Fat," "100% Pure Nature," "100% Real Cheese," "World's #1," "Better Than Fresh Fruit" helps stand out from the competition but could just as easily be noise. Consumers no longer believe false advertising. Honest and transparent product information is what consumers want today.

- **Calories calculation**—Calorie information is usually on the nutrition label. However, to satisfy the calorie-conscious shoppers, highlighting "low calorie" on the front label can be advantageous. But caloric content is not essential information, and the shoppers who pay attention to calories don't represent the majority.

- **Cooking method and time**—"Ready to eat," "2-minute cooking time," "Microwavable," etc. To emphasize how quick and convenient a product is, it is moved up to the front as a product advantage to attract customers. However, cooking method and time are already part of the cooking instruction in the back panel, so this information is repeated.

- **Brand's story, vision, and philosophy**—The story of a granny's most famous recipe becoming a successful, family-owned business; milk coming from the happy cows fed with natural products; a brand's vision of focusing on saving the environment; the corporation's philosophy of believing in innovation of product development, etc.—all of these efforts try to persuade shoppers to trust that this brand is honest, dedicated, and customer-oriented. This additional information could be the 'cherry on the top' of the design or, alternatively, the extra clutters that losses shoppers' attention.

- **Product series introduction**—Listing all product series, sizes, colors, flavors, or related accessories on side panels is another common strategy used for marketing cross promotion. What if there are four colors in this series of product, and they are displayed next to each other on the shelves? Then, mentioning color options becomes repetitive. But what if one highly demanded color is sold out, or this store only carries certain colors? Then, shoppers could walk away with disappointment and shop elsewhere. Instead of having all versions of packaging prints, some manufacturers use one version to fit all to save costs.

- **Use of product**—Another marketing promotion trick is to suggest how the product could be perfect for traveling, parties, camping, picnics, etc. Consumers can use the product as they want, no need for advisement.

- **Recipe**—It's important to provide directions for drug and chemical products. However, cooking instructions for food may not be required but can help improve customer satisfaction. Providing a recipe may not be necessary.

- **Barcodes on all sides**—Putting barcodes on all sides allows cashiers to scan on any side instead of wasting time looking for barcodes. The duplication is unnecessary, though.

Less words

Less really is more when it comes to advertising; the same holds true with packaging. Twitter allows only 280 characters or less for each post. "Four out of five people only read the headline in an advertisement," said David Ogilvy. Keep words short and sweet to allow consumers to quickly grasp any key point. The trend of short scriptwriting from reading on a screen is the same concept as having less words on a packaging label when shopping in store. Fewer, simpler words mean easier understanding for shoppers.

How to simplify the words on packaging?

- Use the words as you say. Use simple words to communicate complex ideas. Avoid using terminology that only you know. Write like you are talking to a new friend. Don't talk down but be welcoming.
- Remove unnecessary words. Use short, punchy sentences and precise keywords. The headline on an article should equal the packaging's front panel. Shoppers don't have enough time to read through your long sentences when browsing the aisles at a grocery store.
- Use informational graphics to replace text. A picture says a thousand words. Infographics can more easily catch audiences' attention and increase brand

Figure 1: Eskay Skincare, courtesy of
Caterina Bianchini Studio

awareness than text. A good icon is like a universal
language that everyone can understand. This can
resolve issues in foreign language translations.

Simple lines and shapes

Minimalism derives from geometric shapes, influenced by
Bauhaus, architect, and product designs. Due to today's
advanced technology, the definition of minimal shapes
and lines remains simple but more organic and freestyle.
Car shapes, for example, are now curvy rather than boxy.

Geometric shapes like squares, rectangles, triangles, circles,
or uniform measurement shapes don't often appear in
nature. Organic shapes are un-straight, un-uniform, uneven,
and un-measurable shapes associated with nature. 'Organic'
now is a term in modern design theory that influences all
design fields including packaging (Figure 1).

Whether designing with geometric or organic shapes and
lines, keep it simple!

Minimal colors

Minimalist design is not about using only black and
white colors; it's about simplifying color shades and
palettes. When 3D technology was introduced to
desktop publishing in the 1980s, graphic designers loved
using gradient and shadow to add the appearance of
dimension. Nowadays, gradient and shadow might mean
cheap or outdated if not carefully simplified. Without

complex shades and messy rainbow color palettes, solid color or single-color objects easily stand out from solid backgrounds without adding any shadows when using the appropriate colors, harmony, or contrast.

The trend of minimizing color is easily seen in logo development. In the standard guidelines of a corporation identity system, it is common for designers to set up standard logos with different versions: one-color, two-color full-color, and B&W. However, designers are no longer stubborn about using all colors, often choosing a single color or even going black or white. These changes are reflected in Apple's logo history.

Look at the three most popular car colors. Different brands and geographies show slightly different results, but based on overall buying behavior, the three colors that never go wrong are silver, white, and black. Though the most minimal color is no color at all (e.g., black and white), that doesn't mean every product or brand needs to be black and white. If all packaging used only black and white, how would anything stand out?

Additionally, not every product is suitable for black and white. The food industry, for example, commonly uses yummy, warm, orange tones—like McDonald's. But for some restaurants and sweet brands, black and white is used to emphasize the theme of luxury. If simplicity is not simple, black and white is not just black and white. Between black and white, there are grey scales. When designers think that black is too plain and boring, they might use charcoal, an off-black, to replace black.

Minimal images

When the strategy of minimal color succeeds, photos become a stand-alone image; illustration and icon become flattened or even eliminated entirely to leave text only.

In minimalist photography, after omitting needless elements, the subject might stand alone. An eye-catching subject on a simple background gets all the attention. Thus, picking a striking and engaging subject is very important.

With the advent of cell phones, AI (Artificial Intelligence), and VR (Virtual Reality), society is flooded with technology. When everything is made by machine, people value a 'human touch' more than ever. Computer lettering can't beat handwriting; machine-made can't beat handmade. Shoppers like unique and personalized items. This makes an organic style illustration with imperfect strokes and simple shading more appealing to viewers.

Icons, symbols, and logos usually are already simple versions of images. The evolution of logo design is getting simpler and simpler. Minimalist symbols in an internet world are desirable.

More negative space

Negative space is the space outside of subject. In minimalist design, the interactions between subject

Figure 2: Feldspar, courtesy of in-house design team

(positive space) and negative space lead the viewer's eye. In Asian art theory, negative space is intentionally used for artistic mood or as a concept. An artist whose artistry lies in utilizing negative space is considered clever and superior (Figure 2).

Leaving a blank background isn't as easy as it looks. Nor does it work on all packaging designs. Some designers feel uncomfortable leaving empty space and reluctantly add unnecessary information to fill the blanks. It becomes chaotic to look at. If the product size is large, then the front panel label could have plenty of negative space and have room to breathe, unlike on smaller labels.

Throwing subjects in the middle is a common page layout. However, center style can be unattractive and doesn't work for every project. Therefore, the most pleasing proportion from art theory's 'Golden Ratio' and from photography's 'Rule of Thirds' are more off-center and lead to more playful and interesting focal points.

5 Case studies of successful and failed designs

Minimalist packaging design may look simple and less involved, but it is not as simple as it looks. In the past two decades, there have been many revolutions in branding and packaging. There are some successful cases, and some failed ones as well. What can we learn from these case studies?

Successful case study:
Puma's Clever Little Bag

- **Background:** Puma is a German multinational company that designs and manufactures sportswear, shoes, and products. San Francisco-based design firm fuseproject partnered with Puma to redesign their shoe packages to be more sustainable.

- **Changes:**

  √ All-in-one: This new design is for both a shoebox and a shopping bag. It's a recycled PET bag that eliminates the need for a usable shopping bag. It's a shoe box, polybag, and a handbag all in one.

  √ No box: Without a traditional shoebox, it uses much less cardboard and could be unfolded to lay totally flat.

  √ No tissue paper: It eliminates useless tissue paper.

  √ No changes: It required no changes to Puma's existing global infrastructure.

- **Fact:**[21]

  √ Puma's Clever Little Bag redesign packaging won 11 international major design awards.

  √ It is made from recycled plastic and reduces cardboard material usage by as much as 65 percent.

  √ Over time, the company will save 8500 tons of paper from its production process.

  √ Its lightweight packaging will help lead to a notable reduction of fuel and carbon emissions in the process.

- **Conclusion:**

  √ No change required in its existing global infrastructure. Shipping, assembly, warehousing, and flow to retail stores needed little to no modifications.

  √ It is a successful, brilliant design that minimizes materials and visual design elements.

Failed case study: Tropicana packaging redesign

- **Background:** Tropicana is a top fruit juice brand, established in 1947. In 2009, new owner PepsiCo decided to redesign the existing packaging design for its bestselling orange juice in the North American market.

- **Changes:**

  ✗ New brand logo: The original and classic curved logo with a bold face and leaf on top had existed for over half a century. It was suddenly changed to a new vertical logo in a sans serif typeface. Vertical text is harder to read, not appealing, and forgettable. The typeface couldn't stand out due to be too similar with the other text. Not only were consumers unable to recognize the Tropicana brand, it was not eye catching.

  ✗ New product images: The original image of an orange with a straw hints at its pure nature. When this orange image was eliminated and replace with a pumped juice, it showed what is literally inside. But what is the juice?

Mango juice? Some yellow liquid? It said "orange," but what if people were unable to read it?

- × **New lid:** The new lid had a special half orange shape to emphasize the squeezable concept. It could have been considered a fancier and more interesting design, but this little improvement didn't score. It was too small to be noticeable from a distance on store shelf.

- **Fact:** A few days after launching the new packaging, consumers began criticizing the new design, especially on social network. Two months later, sales dropped by 20 percent, and Tropicana lost 30 million dollars. Tropicana announced it would return to its original packaging. In total, this initiative cost Tropicana more than 50 million dollars.[22]

- **Conclusion:**

  - ✗ **Identify and value branding assets:** To a big brand with a long history like Tropicana, branding is its biggest asset. When seeing the pink drumming rabbit, consumers know that is the Energizer Battery without seeing its logo or name. After years of spending billions of dollars on commercials, Tropicana's 'straw in an orange' was planted deep in consumers' minds. When redesigning packaging for simplicity, identifying truly valuable brand assets is the first and most important step. After all, the principle of minimalist design is about focusing on what is important. If looking at Starbuck's logo history, you will find that no matter how simplified the design, the two-tailed mermaid never goes away.

✕ **Moderate packaging renovation:** Consumers couldn't recognize the new logo, new typography, new slogan, new lid, or new product image. Changing everything at once confused loyal consumers' buying decision. Aggressively redesigned packaging is risky.

✕ **Simplicity is not simple:** Consumers often view less content, elements, and everything else as having a smaller budget on marketing and design. This type of low-budget packaging is often seen in discount brands.

6 Next minimalist packaging

Environmentally friendly

According to 'The Power of Packaging: 2018 State of Industry Report' by BRAND Packaging, 57 percent of packaging professionals believe that sustainability and recyclability are the top packaging trends. These professionals also look to less BPA, Styrofoam, and plastics materials.[23] Using environmentally friendly packaging becomes a brand's pledge. The back panel of Ice Mountain water packaging says:

Small Cap = Less Plastic

Did you notice this bottle has an eco-slim cap? This is part of our ongoing effort to reduce our impact on the environment. This bottle and cap contain an average of 20% less plastic than our original 500 ml Eco-Shape® bottle and cap. Be green.

Figure 3: Eat Food, Grow Plant, courtesy of Michal Marko, Modest Studio

Dell and Ikea, for example, started using biodegradable mushroom-based packaging to make efforts at reducing waste and increasing recycling. Modest Studio's disposable, eco-friendly carry-out food container, "Eat Food, Grow Plant" (Figure 3), can provide a plant with nutrients and decomposes after a week.

According to the Center of Sustainability & Commerce, the average person in the US generates 4.3 pounds (1.95 kilograms) of disposable waste trash per day. Many innovative minimalist packaging designers make further efforts to provide sustainable solutions by reducing unwanted plastic waste and using reusable and recyclable materials. After learning from the failed packaging redesign project in 2009, Tropicana later opted for a transparent PET bottle in 2012 and even further launched a clear recyclable and reusable 89oz handleware. Consumers could continually reuse this pitcher bottle.

Data from the US Environmental Protection Agency (EPA) indicates that packaging waste declined by using less and lightweight and recyclable packaging materials.[24] In this mass online shopping era, less material and lightweight packaging can not only reduce overhead and selling price but also shopping cost in this competitive and cost-conscious market.

Though going green is the trend, it's still a long, hard road to sustainable packaging. Unfortunately, the recycling process creates far more pollution problems, so commonly

Figure 4: Gauze of Menpu Masda, courtesy of Hidekazu Hirai

Figure 5: Otoño Lberian Ham Packaging, courtesy of Tres Tipos Gráficos

recycled materials cost more than virgin materials. And because of consumers lacking common understanding of sustainable packaging, makers hesitate to invest in changing to environmentally friendly packaging due to inadequate consumer demand.

Keep it transparent

Transparency means honesty to consumers by providing clear information and the real look of products. Consumers are not stupid. Falsely advertised labels only cheat consumers once and results in low returning customer rates. Honesty is the best policy. Shoppers love to see what they are buying. They want to check whether the size is right even though a size chart is provided. They like to feel and touch how soft the fabric is or to see the actual ingredients inside.

Textile manufacturer Maruju designed a cutout window on their towel products allowing shoppers to see the colors and patterns and even touch the textile without opening the packaging (Figure 4). This see-through, peek-a-boo window also works well on Spanish Ibérico Ham to allow consumers to see the balance of fat and meat (Figure 5). Compared to totally sealed packaging blocking consumers from seeing the product until it is purchased, transparent packaging can increase customer's satisfaction.

Okamoto Farm's Sweet Corn Soup was designed with a simple clear plastic bag that allows consumers to see

Figure 6: Oakamoto Farm's Sweet Corn Soup, courtesy of Masayuki Tersahima

Figure 7: Meld's Polyethylene packages, courtesy of Jeannie Burnsid

the actual corn soup. Its beautiful, bright-yellow color contrasts with the black top panel (Figure 6). Meld's see-through ecologically friendly packaging (Figure 7) honestly provides health-conscious consumers its controlled portion sizes and organic foods. Transparent, minimalist packaging is visually simple and informationally straight forward, and also minimizes environmental waste.

Not every product is appropriate for see-through features. Some challenges to transparent packaging include light sensitivity. Light-sensitive food products may be degraded by light exposure. If fully transparent packaging isn't ideal, cutouts or partially covered packaging could be a solution. Manufacturers need to keep visibility in mind. To stay competitive and keep their product advantage, food makers need to adjust recipes to make key ingredient bigger so consumers can easily see it.

Connect to social media

Packaging design is no longer just for retail shelves. Designers also need to ensure the presentation looks good on social media and the tiny screen of e-marketing for mobile devices. Consumers post product pictures to share how much they enjoy certain products on social media. With social media increasingly influencing consumer's buying decisions, having an attractive packaging design that is simple and visually eye-catching becomes more and more important. A Brooklyn-based ice-cream maker

Van Leeuwen had 50 percent increase in sales thanks to its new minimalist packaging redesign.[25]

There are usually two ways of sharing product pictures on social media: through search engines or using your phone to take one. Photos on search engines are usually commercial photography, a standard submission from product manufacturers to e-commerce stores. For easier programming and better web page designs, product pictures are often required to have a fixed size with different views of the front, side, and back. A product beauty shot not only needs a great photographer but also a good model—an excellent packaging design that looks good in the picture!

Banner ad sizes are tiny and hard to view on mobile devices. The smaller the picture, the smaller the area of the packaging label will show. Less clutter keeps it simple and easily seen in tiny sizes. Minimalist packaging design in social media is more attractive and efficient.

7 Conclusions and challenges

For throwing off the shackled images of traditional ostentation, corporations strive for minimalist packaging design to be contemporary and to uplift the value of products and brand images. Keeping the principle of omitting unnecessary things and carefully evaluating every element and word to serve its essential purpose, minimalist packaging leads to cleaner, sharper, and classier looks. Packaging's attractive sophistication acts as a silent salesman that influences consumers' buying decisions and reaches its ultimate goal of reducing costs and promoting sales. Minimalist packaging is on trend now and not expected go away in the next decade.

However, is minimalist design suitable for all packaging? According to a Luminer's survey, 60 percent of consumers are likely to reject a product if the label does not provide enough information about the product, and 37 percent are deterred by inadequate photos and small text on the label. Too much information is cluttered; however, over-simplified packaging can scare away customers due to insufficient product information. Minimalist packaging looks deceptively simple. The true question is: What is enough? Unfortunately, there is no one answer that fits all. Different brands, product histories, and industry types must be factored in before deciding to go for a minimalist design.

Designers are constantly searching for solutions to resolve all packaging challenges with each passing year. Technological innovation, material improvement, and shifts in consumer buying behaviors constantly influence minimalist packaging movements. Tomorrow, there will be more and more sustainable packaging materials for packaging designers to choose from; there will be enhanced solutions for keeping product transparent that won't rot from exposure; there will be the next social media that connects consumers.

Are you ready for your next minimalist packaging project?

Notes

1. Schonlau, J., *Minimalist Graphics*, (New York, Happer Design and Maomao Publications, 2011).

2. Forbes.com, '2018 Ranking: The World's Most Valuable Brands,' (retrieved from: The World's Most Valuable Brands: https://www.forbes.com/powerful-brands/list/, 2018).

3. Segall, K., *Insanely Simple: The Obsession That Drives Apple's Success*, (New York, Penguin Group, 2013).

4. Richard, S., 'Jonathan Ive Interview: Simplicity isn't Simple,' (retrieved from: https://www.telegraph.co.uk/technology/apple/9283706/Jonathan-Ive-interview-simplicity-isnt-simple.html/, 2012).

5. Kilchii, D., 'Apple is No Longer a Consumer Brand—It's a Fashion Icon,' (retrieved from A Medum Corporation: https://medium.com/@DraketheFox/apple-is-no-longer-a-consumer-brand-it-s-a-fashion-icon-317b4f0b06f2/, n.d.).

6. 'Supermarket and other grocery store sales in the United States from 1992 to 2017 (in billion U.S. dollars),' (retrieved from: https://www.statista.com/statistics/197626/annual-supermarket-and-other-grocery-store-sales-in-the-us-since-1992/, 2019).

7. 'FMCG and Retail: Looking for achieve new product success?,' (retrieved from: https://www.nielsen.com/us/en/insights/reports/2015/looking-to-achieve-new-product-success.html/, 2015).

8. 'How does society influence art? And can art change society?,' (retrieved from: https://www.quora.com/How-does-society-influence-art-And-can-art-change-society/, 2017).

9. Nielsen, J., 'How little do users read?' (retrieved from Nielsen Norman Group: https://www.nngroup.com/articles/how-little-do-users-read/, 2008).

10. Lupus, M., '9 Inspiring Packaging Design Trend for 2019,' (retrieved from: https://99designs.com/blog/trends/packaging-design-trends-2019/, 2018, Dec).

11. Loredana Papp-Dinea, M. B., 'Behance: Trend of the year 2019,' (retrieved from: https://www.behance.net/gallery/71481981/2019-Design-Trends-Guide/, 2018, Oct).

12. Roberge, D., 'A Predictive Look Ahead: 2019 Product Packaging Trends,' (retrieved from: https://www.industrialpackaging.com/blog/a-predictive-look-ahead-2019-product-packaging-trends/, 2018, Nov).

13. Nemel, T., '10 Creative Packaging Design Trends for 2019,' (retrieved from: https://inkbotdesign.com/packaging-design-trends/, 2018, Oct).

14. 'New Year, New Packaging Design Trend,' (retrieved from: http://sourcenutra.com/2018-2019-packaging-trends/, 2018, Dec).

15. Crowd, C., 'Creative Packaging Design Trends for New Products Moving into 2019,' (retrieved from: https://www.cadcrowd.com/blog/creative-packaging-design-trends-for-new-products/, 2018, Sep).

16. 'Shopper reveal the packaging & labeling that grab their attention,' (retrieved from: http://www.luminer.com/articles/survey-packaging-labeling-grab-shoppers-attention/, 2018).

17. Joker, K., 'The Power of Packaging: 2018 State of Industry Report,' (retrieved from: https://www.brandpackaging.com/articles/86137-the-power-of-packaging-2018-state-of-industry-report?v=preview/, 2018).

18. 'NPR/Marist Poll: Amazon is a colossus in a nation of shoppers,' (retrieved from: https://www.npr.org/about-npr/617470695/npr-marist-poll-amazon-is-a-colossus-in-a-nation-of-shoppers/, 2018).

19. 'Understanding the Power of a Brand Name,' (retrieved from: https://www.nielsen.com/us/en/insights/news/2015/understanding-the-power-of-a-brand-name.html/, 2015).

20. 'Shopper reveal the packaging & labeling that grab their attention,' (retrieved from: http://www.luminer.com/articles/survey-packaging-labeling-grab-shoppers-attention/, 2018).

21. 'Puma Clever Little Bag,' (retrieved from: https://fuseproject.com/work/puma/clever-little-bag/?focus=overview/, 2011).

22. Marion, 'What to Learn From Tropicana's Packaging Redesign Failure?' *The Branding Journal,* (retrieved from: https://www.thebrandingjournal.com/2015/05/what-to-learn-from-tropicanas-packaging-redesign-failure/, https://www.nytimes.com/2009/02/23/business/media/23adcol.html?pagewanted=all/, 2015).

23. Lempert, P., 'Ikea Switches to Packaging Made From Mushrooms,' (retrieved from: https://www.winsightgrocerybusiness.com/retailers/ikea-switches-packaging-made-mushrooms/, 2018, June)

24. Lilienfield, B., 'Live Science: From Crisis to Myth: The Packaging Waste Problem,' (retrieved from: https://www.livescience.com/50581-packaging-no-longer-the-nightmare-some-claim.html/, 2015).

25. Quito, A., 'A Brooklyn ice cream brand increased sales by 50% after it redesigned its packaging,' (retrieved from: https://qz.com/944549/van-leeuwen-ice-cream-sales-increased-after-it-redesigned-its-packaging/, 2017).

Case Studies

· · · · · · ·

PART 1

RESTRAINT

Focus on what is important and disregard unnecessary content.

"Minimalism is not a lack of something. It's simply the perfect amount of something."—Nicholas Buroughs

What is the perfect amount? Different industries have their own answers and standards. Keep it simple!

Les Bons Vivants

Design

Cécile Nollier, Clémence Gouy

Completion

2018

Les Bons Vivants is a mobile delicatessen offering assortments of gourmet food that can be discovered in a special box every month, starting with the most French products you can think of: cheese and wine.

Packagings in supermarket are often very busy. By going for a clean, minimal box, the designers wanted to mark the distinction between Les Bons Vivants' cheese and the established cheese products out there. Clean and minimal, the packaging doesn't tell too much and allows the product to speak for itself. The bright klein blue catches the attention while still reminding consumers of the traditional blue color used for milked products in France. The minimalism also sparks curiosity, making customers want to see what's inside.

The design was inspired by the traditional way of packaging local groceries in the countryside: they were often made of a plain kraft wrapping paper, sometimes with a specific pattern and handwritten labels.

However, as Les Bons Vivants sought to target a young urban population, the designers wanted to give the packaging an updated and appealing look to inspire curiosity. Each pattern inside of the box was inspired by a different cheese texture, and the precious box offers a unique way to discover cheese and wine.

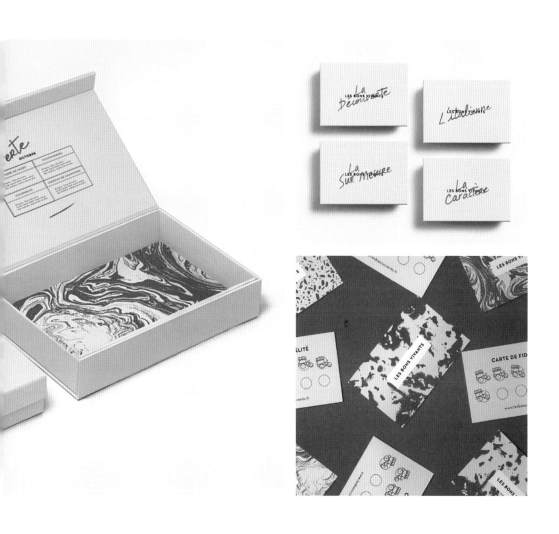

Sundaze

—

Client
Sundaze Skincare

Design
Caterina Bianchini Studio

Completion
2017

Sundaze is a modern skincare brand focusing on sunscreens, which was born from the sun-soaked streets of San Francisco, California. Sundaze hoped to modernize the market through bold design choices and a new wave of organic sunscreen, creating a brightening and moisturizing effect.

The brand identity was designed to explore the use of unique graphic placement. The logotype was developed to sit like a partial eclipse, giving a subliminal nod towards the products as a daily sunscreen. With a simple but quirky design and a flexible logo, this brand offers a unique presence that has personable qualities.

The placement of the word "Sundaze" looks like sun rays, creating a logo that completely references the sun. The design was continued across the packaging, creating a coherent and bold range that sets itself apart from other options in the market.

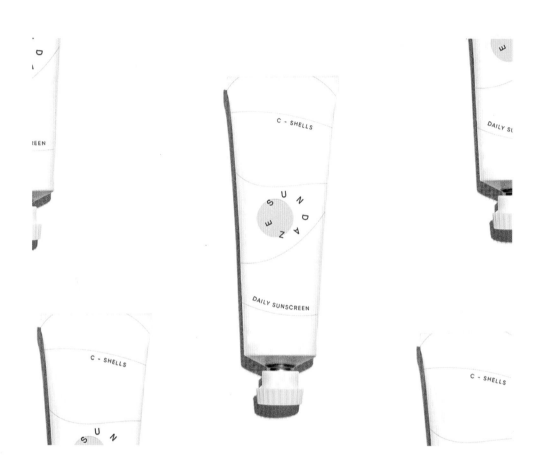

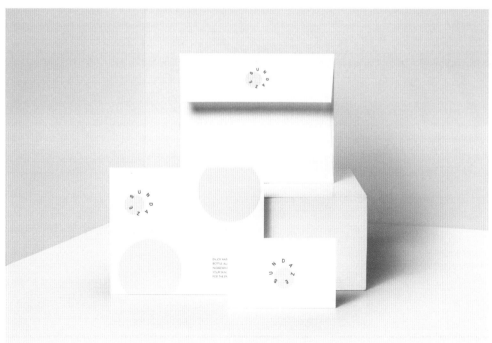

Dose Skin

Client
Niche Skin Labs

Design
Tato Studio

Completion
2018

Located in Canada, Dose Skin started out as a line of skincare boosters dedicated to protecting and revitalizing sensitive skin. Dose Skin uses a refill system with eco-friendly glass vials and aluminum canisters to reduce their environmental footprint. Their objective was to develop a sophisticated skincare brand that combines botanical actives with cutting-edge skincare science.

The packaging was designed to enhance the concentrated formula with the healing attributes. Reflecting the brand's elegance and purity, the designers employed grayscale colors with a powerful symbol that communicates the extremely gentle and powerful ingredients.

The designers tested out different kinds of paper to provide a pleasant tactile and visual experience, and also decided to enhance the minimalistic design by using three-dimensional emboss and foil stamping techniques, taking care of the entire design process up to the production to be sure about each and every detail.

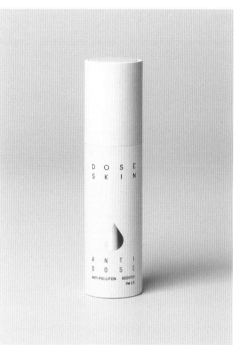

Onkel Toms bottle labeling

Client
Onkel Toms

Design
Michael Reichen

Completion
2018

Onkel Toms is a small craft beer brewery in Switzerland. It produces a wide range of craft beers, rarely brewing the same type of beer a second time. Because of limited financial and human resources, the bottle labeling had to be cost-efficient and simple to execute. It was important to have a simple and flexible concept so that a label could be ready in a short time. In addition, it needed to be someone uniform across all of the beers to be easily recognizable as an Onkel Toms, but at the same time different from each other to underline the variety and different tastes. Black and white was selected to save on costs. The designs are easy to print and cut and can be mounted in a simple way with a stapler.

Feldspar packaging

Client
Feldspar

Design
Feldspar

Completion
2016

This packaging series was for a homeware range. With the brand's philosophy of "objects for life"— making beautiful objects that will last a lifetime— an antidote to a throwaway consumer culture, the boxes needed to reflect this. As such, they were designed to be quietly elegant, not shouting to be seen, pared back to just two elements: beautiful high-quality paper in muted colors and only essential information embossed onto the boxes.

The brand focused on using the best materials that could be found, and the designers carried this through into the packaging. The goal was to create a box that customers would want to keep and re-use for storage or even to have on show.

The designers were very inspired by their experiences in Japan and the beautiful care that is taken there with purchases in regard to wrapping and packaging, even just a pair of socks. As Feldspar's homewares are made to last a lifetime, the designers wanted the act of buying or receiving one of their products to be special.

Everything they make comes in its own gift box made from paper that is milled in the Lake District and embossed with matt black and gold lettering, and then hand-assembled in Malvern (Worcestershire, England). The boxes are left un-coated so the texture of the paper is central to the experience, while the colors (a different one for each size) reflect the landscape around the studio in Devon (England), reminiscent of mist on the moor.

CARPOS

—

Client
Carpos Loannis Stoliaros

Design
Panos Tsakiris

Completion
2017

The task was to design a system of products—a bottle and its packaging—that could safely hold the client's olive oil. Located in Switzerland, their main market was the local population.

The chief aim was for the user to enjoy the product with more than one sense and, for this reason, profound textures were subtly incorporated into the design. The inspiration came mainly from the form of an olive-tree trunk interpreted in different ways and applications. The ethos of CARPOS® and the project's guidelines were the tools that guided the design of the limited-edition, numbered, and exclusively handcrafted bottles containing the carefully preserved oil of 750 olive trees from the mountains of Rhodope, Greece.

Each of these bottles was prepared to be ready to be shipped individually in its own eco-sustainable packaging, structured with no adhesives. The minimalistic organic form of the bottle, the materials, and the time-consuming construction were what made it stand out, while the attention to detail and the company's love for what they do were part of the brand value.

Part of the design process was to achieve minimalism as part of the core value of the brand to communicate the vision and overall personality to the end user. If the message is powerful then the packaging needs nothing more than the essentials, which the designer believed to have achieved with this project.

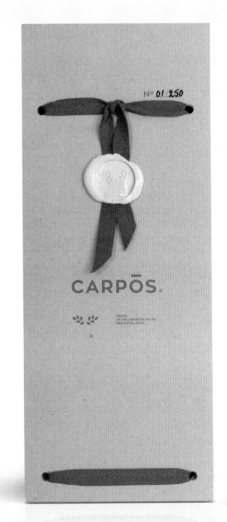

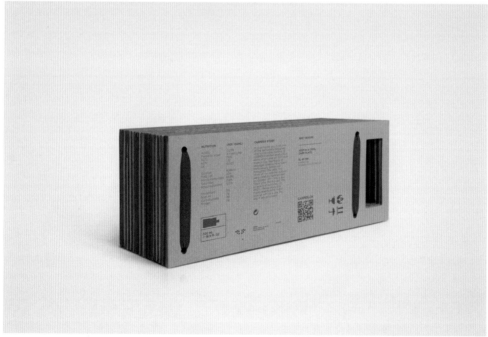

Eumelia Olive Oil packaging and branding

—

Client
Eumelia

Design
Typical Organization

Completion
2016

The creation of the visual brand identity and packaging design for Eumelia Olive Oil commenced with the Eumelia color theory. A seemingly inappropriate color choice for the Eumelia biodynamic products, the earthly red/rose has long been identified—by Goethe's color scheme from 1772—as a complement to the green of the earth.

This was further bolstered in the 20th century by Rudolf Steiner—the founder of biodynamic farming—who claimed red/rose is the color of the inside of the womb. Hence, the color became the branding solution for Eumelia, in itself depicting the precious liquid of olive oil delivered by mother nature herself.

Rudolf Steiner was also active as an architect with a particular approach of completely rejecting the use of the right angle in design and replacing it with curves and smoother slopes observed in nature. Applying this principle to the olive oil package with a Helvetica Neue Medium font, the designer succeeded in promoting the biodynamic range and also winning the EBGE Design Award 2017.

Steph Weiss

—

Client
AND UNION

Design
AND UNION

Completion
2016

The first Steph Weiss canned beers were unveiled at the beginning of 2017 with a bold yet restrained design. Solid, single colors representing each beer in the line-up were created in an attempt to cut through the clutter commonly seen on shelves these days.

The modest hero of the design is texture, as the designers played with embossed geometric shapes to add depth to the container in the absence of typical craft beer aesthetics. By manipulating the surface of the aluminum can, the design balanced visual simplicity with detailed, tactile consideration to reflect the malleable nature of the material.

The minimalist design echoed AND UNION's modernist ideals and was a deliberate rebellion against the tired idea that cans are for cheap beers.

Bodega Los Cedros packaging

—

Client
Bodega Los Cedros

Design
Anagrama

Completion
2017

Bodega Los Cedros represents a Mexican vineyard located in the mountains of Arteaga, Coahuila, which has been dedicated to the wine harvest since 2012.

Based on the geographical location of the mountains where the vineyard is located, this brand solution highlighted characteristic components of the area such as climate, altitude, flora, and fauna. The logo employs three pine trees as the brand's distinctive icon.

The wine label and die-line mimic cloud shapes. The sans serif typographic selection adds modernity while the script gesture preserves a more organic touch, accomplishing the brand's contrast between simplicity and elegance.

The designers included a neutral color selection to highlight the diverse wine color tones and to accentuate the contrast between the bottle and label, while the packaging composition recalls the landscapes around the winery.

Norske Bryggerier identity

Client

Norske Bryggerier

Design

Frank Kommunikasjon

Completion

2018

Norske Bryggerier is an Oslo-based company that aspired to start up small local breweries along the west coast of Norway and develop a portfolio of high-quality craft beer brands.

The concept behind the identity was all about the taste of quality beer and good company. The packaging created was a "non-sale" design used only for internal tasting for quality control before full production, once the testing is deemed successful. Taking great pride in tasting their products against the competition, Norske Bryggerier knows that with so many quality breweries in Norway and internationally, the game is getting harder.

The design was inspired by pharmacy bottles where the label is purely for a practical card file system. The reason for the neutral design was to avoid being affected by colors and illustrations, which can lead to false expectations, and to focus purely on the tasting process. The "N" stands for "Norske" and the two circles have the double meaning of both a colon, which represents many breweries, and two glasses, which signifies the company of friends—the best way to enjoy their beer. The black represents the "dark market" in Norway, which stands for the law that prohibits the advertising of alcohol or use of promotion of any kind to generate sales.

Strike Wine

Client
Conrado Gajate / Kiko
Calvo

Design
Javier Garduño Estudio de
Diseño

Completion
2017

The client asked the designers to work on the naming, branding, and packaging for their wine. In the first meeting, the client gave the designers complete creative freedom, describing the wine as full. In Spanish, the word for full is "pleno" and it has different meanings. One of them is related to bowling and means "strike," as in, victory. So the designers decided the bottle should be a bowling pin. In order to give it this look, it has a printed sleeve and it's sealed with the synthetic white wax seal. The outcome is a visual metaphor that will stand out of every shelf, suggesting "Let's play!"

Mandarin natural Chocolate

Client
Mandarin natural Chocolate

Design
Yuta Takahashi

Completion
2016

Mandarin natural Chocolate makes "bean to bar" chocolates, employing chocolatiers consistently working on the entire manufacturing process from the roasting of cacao beans to the finished chocolate bar. The designers offer 60, 80, and 100 percent cacao chocolate bars made only with organic cacao and cane sugar.

The designers wanted to express the idealism of this chocolate brand in the package design. To express that the chocolate was organic, without impurities, pure and of good quality, the designers used a white packaging to give the impression of exclusivity. The dots positioned on the lower part of the packaging symbolize the cocoa blending ratio of the chocolate in a simple manner. And the logotype and dots were positioned such that they are on opposite ends of the packaging to further bring out the whiteness in the center of the packaging, through which the designers hoped to add on to the impression of minimalism.

Mandarin n...

Mandarin n...

Mandarin natural Chocolate

A black-and-white color scheme tends to appear sophisticated, formal, elegant, exclusive, and prestigious.

With the elimination of needless colors, black and white offers a powerful palette in minimalist designs. The high contrast of a pop of black on a white background (or vice versa) increases the visual impact of packaging graphics and creates an interesting surprise that is both unique yet familiar.

Just as "black tie" is a formal style of dress and "black belt" symbolizes an expert level of ability in martial arts, in design a black-and-white palette is often representative of high-quality products.

5:min

Design

Jessica Wonomihardjo

Completion

2016

Skincare can be a difficult area of sales, especially for men, so creating a timeless, straightforward, and functional design was considered crucial in this project.

Time and the form of a clock were the primary metaphors that drove the overall brand identity and package design. The shape of the bottle mimics the shape of a traditional clock. The concept of time could also be applied to how the consumer uses the skincare product. As such, the product represents not only an effective skin treatment, but also time efficiency.

The Avenir typeface was used to express the contemporary and purist values that the brand upholds. Monochromatic colors—black, white, and gray—were used to ensure brand longevity.

The fundamental concept behind this product was to be straightforward and functional. Having the main emphasis on the order of use and product name works well with the consumer, while the use of black and white enhances the classy and sophisticated look of the brand.

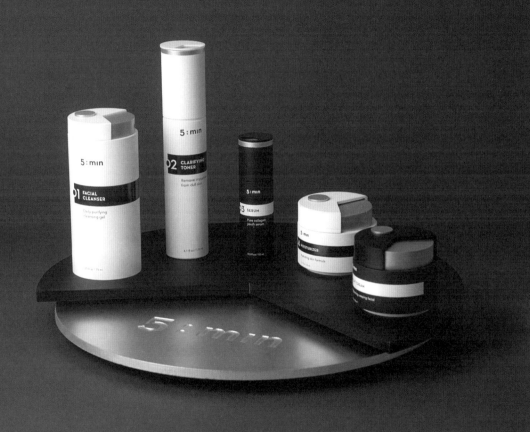

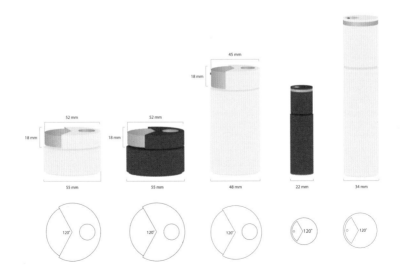

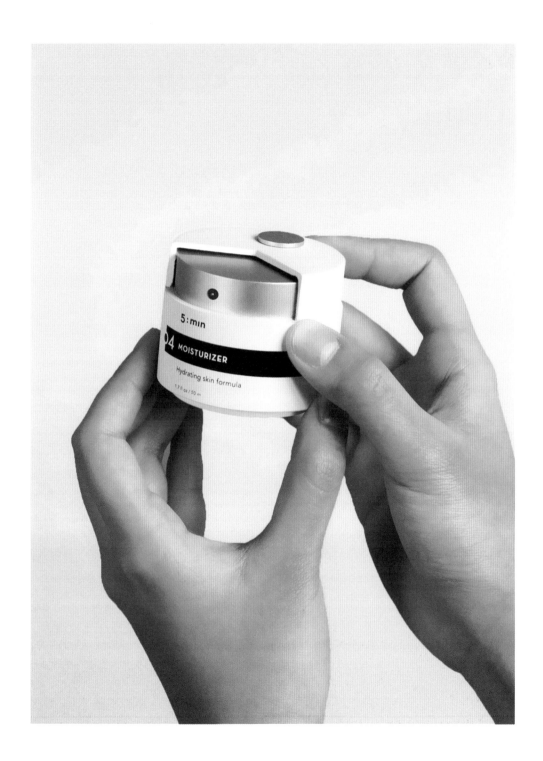

Peet Rivko packaging identity

—

Client
Peet Rivko

Design
Gunter Piekarski

Completion
2017

In response to her personal struggle with sensitive skin, the client launched Peet Rivko, a line of clean, plant-based products. Since commencement, the Brooklyn startup has garnered much press attention for its all-natural approach and clean visual aesthetic.

This packaging design responded to the brand's "skincare, simplified" message with a reductional visual language. Pared-back production, playful typography, and a clean graphic system were translated into an honest, dynamic, and uplifting packaging.

K&Q Chess Stick Cake packaging

—

Client
Poco a Poco

Design
Latona Marketing Inc.

Completion
2017

Stick-shaped sweets (cakes, baked goods, and chocolates) are often sold wrapped in clear plastic. However, because of their long, thin shape and the unstable plastic packaging, traditionally there isn't much usable space for design elements, and those present are often difficult to read.

The design of K&Q Chess Stick Cake packaging solved these issues by covering the plastic bag in an embellished paper sleeve. With a length-to-width ratio of 8:1, the sides of the sleeves are quite long and are covered in a checkerboard pattern. When all eight sleeves contained in this gift set are aligned, the beautiful checkered pattern of a chessboard is revealed. The pattern continues onto the front, which also features a centrally located window through which the contents of the sleeve can be seen.

These paper sleeves were designed to keep costs to a minimum with no glue required for their assembly. Each sleeve is made out of a single piece of die-cut paper that allows the final shape to be formed by hand in only one step.

MARAIS Piano Cake packaging

—

Client
Patisserie Perle

Design
Latona Marketing Inc.

Completion
2016

Usually, cake gift boxes simply line up all the goods neatly. However, MARAIS Piano Cake took a different approach with their boxes of individually wrapped cakes.

The inspiration for this design, the piano, came from the idea that a piano is an instrument that creates beautiful harmonies from the combination of each individual key's sound. Patisserie Perle felt that society is a lot like this as well. Like the keys of a piano, where no two keys have the same sound, we all have our own individuality. Similarly, like the keys of a piano, we are all of different colors and creeds, and yet we come together to create a beautiful music.

In order to stick to the budget, the designers focused on only one design using all six surfaces to recreate a piano keyboard, which can be adapted in size—from small keyboards to full 88-key grand pianos and even larger. For example, for one octave of 13 keys, they use eight cakes. Two octaves, or 25 keys, would be 15 cakes.

A gift-style product, the client sought to create something that was made with care. The designers crafted the gift boxes with this in mind.

Dietary Supplement—White Sorbent Extra

Client
Bioterra

Design
Lesha Limonov

Completion
2018

The design concept of this packaging was inspired by the domino principle. To achieve a goal, you need to start from something small. The first step will cause a chain reaction of a new one, which then provokes another, and so on in a linear sequence, according to the domino principle. The same is true of our health: it's enough to start with the first pill.

The packaging of Dietary Supplement—White Sorbent Extra was crafted in the form of dominoes with a pattern 6 | 4. The white circles are associated with white tablets, which were embossed just like traditional dominoes. The name was made by foil stamping. The number of tablets in the package is the same as that on the front side of the pack. On the reverse side is the duplication of the name in Braille.

The designer tried to use non-standard ideas and solutions when creating the package. It is important to understand the essence of the object and to find this very idea through associations. When the idea is there, the design becomes only a technical point. The designer used the principle of metaphor and the result was vibrant and gave a new context.

Tiger Beer Air-Ink

Client
Heineken Asia Pacific

Design
Marcel Sydney

Completion
2017

The Air-Ink™ product is made through innovative technology that collects emissions directly from vehicle exhausts, transforming air pollution into safe, reliable ink. Thus, the packaging for this beautifully simple idea and tool for creative expression needed little adornment, decoration, or mediation. A crisp, considered, and restrained packaging form reflected the detailed engineering and attention to construction. On to this stark, blank canvas, the designers applied an ink blot graphic device and elegantly playful arrangement of typography, deliberately avoiding bright colors in favor of bold, high-contrast black and white.

The design inspiration was the product itself, a breakthrough technology that transforms air pollution into a tool for producing works of beauty. Air pollution settles on the street as black, carbon soot, thus the designers stayed close to a black-and-white aesthetic, seeking to reflect both the transformative technology and the deeply pigmented black of the ink itself. Elegantly playful typography in all capitals formed a severe visual architecture, contrasted with the organic, fluid forms of a richly textured ink-blot graphic device.

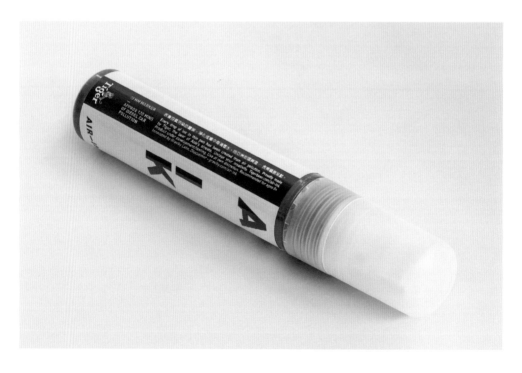

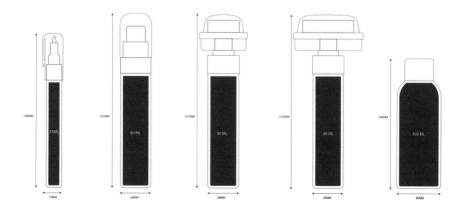

Soto Sake

Client
Billy Melnyk

Design
Joe Doucet x Partners

Completion
2017

After reviewing hundreds of names, the designers all settled on Soto, which means "outside" in Japanese. It seemed appropriate for countless reasons, but most pointedly as the designers were bringing the very best of this spirit at the heart of Japanese culture to the west in a way that had never done before. The idea of a hole through the bottle was a direct offshoot of the name, allowing one to view the "outside" world through the sake. The designers chose to keep Japanese characters on the bottle to reflect the fact that this wasn't a western sake, but a super-premium sake brewed in the traditional methods in the historic Niigata region.

One hurdle the designers faced was that the bottling facility in Kyoto wasn't equipped to insert corks in the sake, only the screw cap found on most sake bottles. This was an issue as most westerners see a screw cap as "cheap," yet this product was one of the finest Junmai Daigingo (the highest level of sake) ever produced under $50. The solution was to cover the cap. Billy suggested Japanese denim, which was a stroke of genius. The designers chose to train the sale teams to teach a ritual when taking off the cloth topper—using it to wipe down the inevitable condensation that would appear on the chilled bottle, and then setting the cloth on the table and presenting the bottle on top of it. This bit of sleight-of-hand completely distracted from the screw cap and turned a potential liability into a powerful strength.

Eternal Sunglasses

—

Client
Roswood

Design
F33

Completion
2016

Creators of high-quality sunglasses made from traditional materials yet at an inexpensive price and unlimited, free replacements for the life of the glasses, the designers sought to craft a packaging that could surprise and also contribute value to the product itself.

In order to express this unique concept, Roswood appropriated the idea of a coffin package for their wooden sunglasses. This striking design executed by F33 creative agency was crafted in the shape of a small coffin with the slogan, "the key to immortality is first living a life worth remembering."

The designers adapted the naming to the shape of the box itself, breaking it into a cross shape and decorating it with something that could look like nails. The result was a modern coffin, very clean and exempt from the negative connotations of death, but with the positive effects of resisting almost all of eternity.

PART 3

MONOCHROME

Minimalist packaging minimizes color palettes, gradients, and shadows to leave only one solid color or analogous color(s).

The color palette of minimalist packaging can go beyond black and white. It can also be balanced with darker and lighter monochrome.

Atölye

—

Client
Atölye

Design
Mustafa Akülker

Completion
2018

With the minimalist approach, it's essential that the brand is easily recognizable and expressed in a simple and effective way. Additionally, brands using minimal designs should be easily rememberable and reflect the brand's purpose.

Atölye—which was established in Nisantasi, Istanbul—sought to expand their range of handmade decorative products with a candle collection inspired by nature. Using fresh colors and a gradient to provide a minimalist aesthetic, the designer sought to reflect the brand within the packaging design. With a focus on the mystery of nature and natural colors, he was able to craft an elegant and simple packaging. The result was a contemporary interpretation in packaging of the age-old product to appeal to the target audience.

Forno Classico rebranding

Client
Forno Classico

Design
Artware

Completion
2018

Forno Classico is a bakery brand in Kavala, Greece. Artware accepted the challenge of re-establishing the design of Forno Classico's corporate identity by introducing the philosophy of designing on a fresh pattern of clear lines. Thus, the new logo was inspired by the classical simplicity of its form and decorated with a stylish handwritten statement of "Classico."

The new identity was integrated into the new package design and combined with a new color palette in orange gradients. With a large set of different packaging applications, dedicated to each and every product, the package itself narrates in an abstract way the product's storyline from production to consumption and pleasure: figuratively, by following the line on the surface of the packaging and, literally, with the printed mottos inside the packaging.

The final touch of the redesign was applied to both business attire and communication material, providing a homogeneous application through all of the brand's references.

Moodcast Fragrance Co. brand identity

—

Design

Studio L'ami
(Fredericus L'Ami)

Completion

2018

In the early conceptual stages of the project, the designer referenced the work of Dutch visual artist Berndnaut Smilde, who in 2012 created a series of self-made clouds in beautiful spaces. The abstract nature of this concept ultimately led to the use of the circle to simulate the effect of the Moodcast candle.

The gold-foiled circle was the central design element intended to communicate one's personal space or atmosphere. The circle was executed in a multitude of ways throughout the product experience. For the glassware, the designer used individually colored matte-glass vessels with a translucent circular window to reinforce the concept of personal space and highlight the flame as it burns. Colors were inspired by 1960s Finnish glassware and access the principles of color-therapy to create a palette that helps reinforce each mood.

A minimal aesthetic demanded that the designer looked for ways to be strategic about how the product information was revealed. Details (weight and volume information) were blind embossed to ensure they're observed on close inspection rather than adding to the overall brand impression.

Giving individual design elements breathing room worked two-fold: the overall impression to the customer is one of calm and each element has more prominence so they can communicate more effectively.

RINGANA packaging

—

Client
RINGANA GmbH

Design
Moodley Brand Identity

Completion
2016

RINGANA produces fresh cosmetics and superfoods that they describe as consisting of one ingredient: nature. Their standard is always as high as absolute freshness goes—without artificial ingredients, without compromises. Their philosophy in regard to the packaging design was that it should transport while it focuses on the essential and also reflect the spirit of innovation.

During the general rebranding process, Moodley put the product group of the supplements into a new packaging too. The first step was to create a reduced, timeless design, which gave an overview view but also enabled consumers to differentiate between the products within one united brand.

Step two was to make an easier box design for handling for customers and employees who prepare products for distribution. The overall result was a clean and fresh packaging design made from 100 percent recyclable materials without any unnecessary razzle-dazzle or decoration.

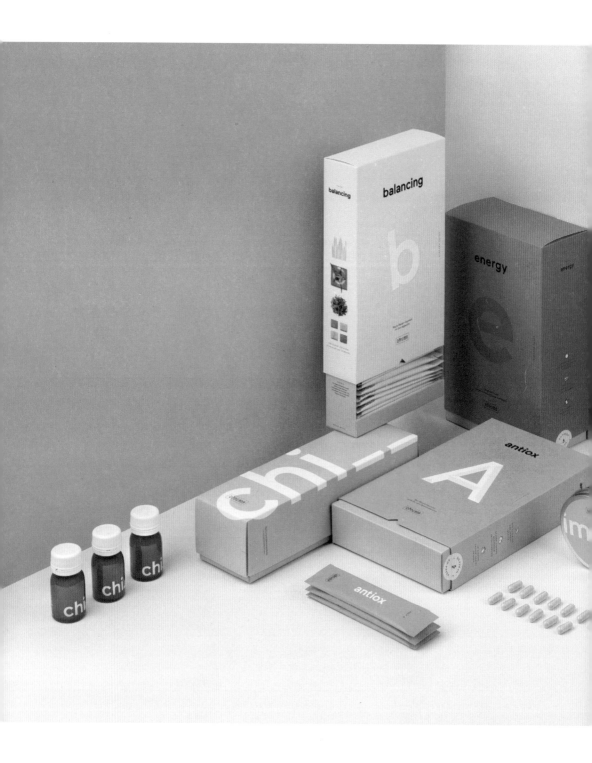

JUS·Juice Up Saigon

Client
JUS·Juice Up Saigon

Design
Duy—N

Completion
2015

The mission for this packaging design was in researching and finding the right solution both in terms of visual identity as well as art direction to create a unique, sustainable, cold-pressed juice brand in Vietnam. Since cold-pressed juice is a completely new product in Vietnam, JUS·Juice Up Saigon was the pioneer to break the barrier.

Based on the concept of an equilateral triangle of yoga that connects body, spirit, and mind together, the designer created a unique equilateral triangle bottle shape. The JUS logo was designed on simple terms with organic curves to create a natural visual identity for the brand reflecting its organic ingredients.

The design process was unusual with JUS, taking the designers one month to craft and perfect the bottle shape by using clay 3D-printed bottles for testing the right angle for the tricky triangle bottle, with the bottle being produced only after the design was perfected.

After the bottle was produced and the brand was completed, the printing test was another task challenge, as in Vietnam, it is vital to have the right ink formula for plastic or else the print can be peeled off or scratched out easily.

Finally, due to the hot climate in Saigon, heat protection inside the delivery box needed to be designed functionally and produced within budget.

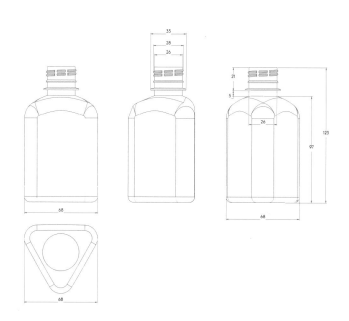

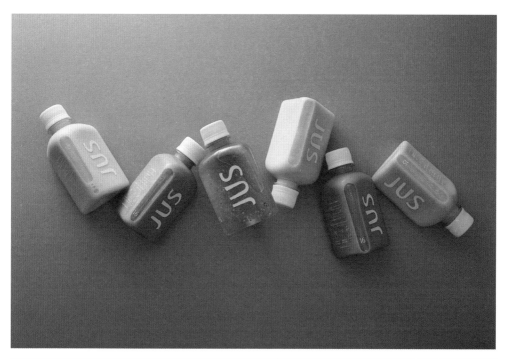

One Power Bank branding and package

—

Client
3R Memory

Design
Veronika Levitskaya

Completion
2017

One is a small boost battery developed especially for urban citizens who do not want to carry around huge power banks. The designer was tasked with creating the logo, identity, and packaging for this tiny power bank. Just like the gadget itself, the packaging was created to be very compact and convenient, and it can be opened in a fast and easy manner. Additionally, the oval shape of the product resembles the letter "O," which is the key element in the identity. The beauty of the design makes it ideal for both personal use and gift-giving.

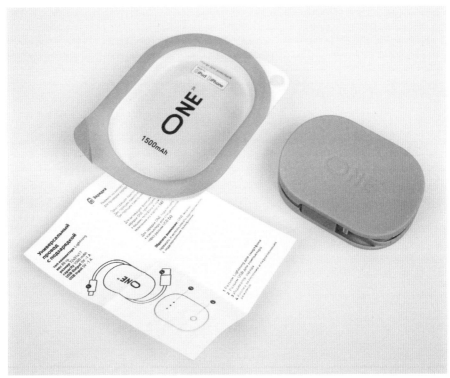

Rubedo Hot Sauce

—

Client
Rubedo Hot Sauce

Design
Stefan Andries

Completion
2017

Claimed to be a hot sauce like no other, based on a secret recipe using home-grown chili peppers and "some alchemy," the designer of Rudedo wanted to recreate the bottle used by alchemists in the Middle Ages for this hot-sauce packaging, but with a completely modern, simplified design. The shape and texture of the bottle, with the bright red color of the sauce and the natural feel of the cork, created the perfect combination for the label to stand out in a simple but unique way.

The Rubedo project—which was named after the Latin word for "redness"—started with the selection of the packaging. Only after crafting the perfect bottle with its long neck, round base and frosted glass did the designer move on to create the logo.

The logo was inspired by the simplicity of the alchemical symbols used in the Middle Ages with the "R" for "Rubedo" looped into the design. Modern and minimalist, the logo also incorporated the ancient symbols used by alchemists to label their potions, which mean water, fire, sun, and earth. "Hot Sauce" and "Rubedo" clearly stand out on the protruding tag for quick identification, with the alchemical symbols on the opposite side to add character, further echoing the alchemist theme.

Teastories branding and packaging

—

Client
Teastories

Design
Anagrama

Completion
2014

Teastories is a tea store located in Vienna offering selected premium tea products. The store offers a wide variety of teas that come from all over the world, from Sri Lankan black tea to green tea from Japan, as well as various tea accessories.

The designers created a brand that alludes to the natural accents of flavor and aroma in tea, yet presented in a minimalistic and approachable way. This was obtained by using brush strokes and a subtle color palette referencing the essence of tea, which was presented throughout the brand's packaging.

The logo brings together quotation marks and tea leaves, playing with the idea that every tea shares a different story. On the tea packaging you can find quotations that describe each tea's personality, thus accentuating the different characteristics in a unique way.

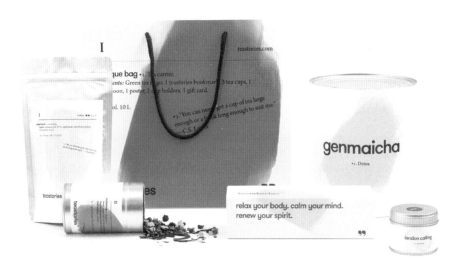

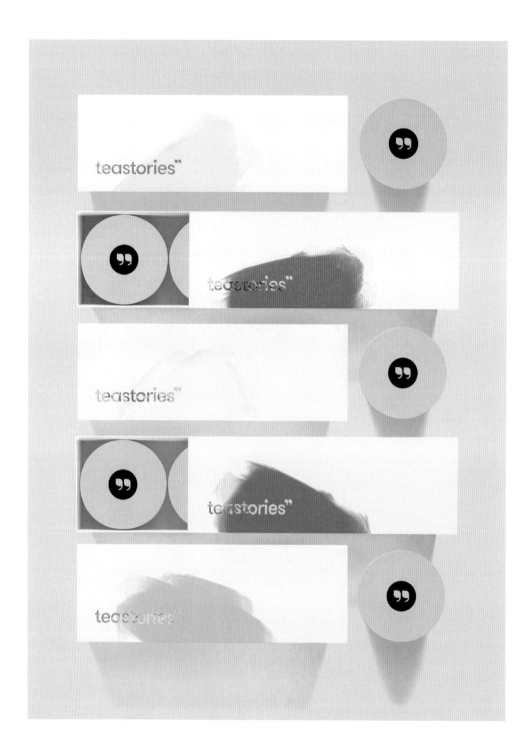

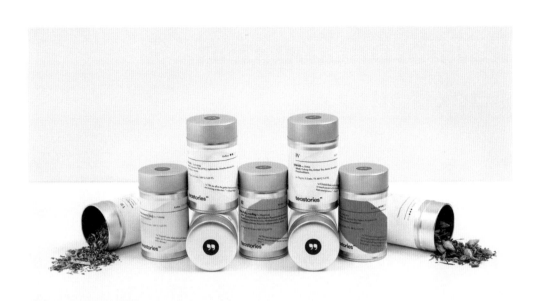

energize your work, life and spirit.
you are amazing and full of energy.

Tesis branding and packaging

—

Client

Tesis

Design

Anagrama

Completion

2016

Tesis was born in Mexico under the curatorship of Sommelier Marcela Garza. Its unique herbal and tea blends were inspired by the origins of tea and ancestral heritage from flowers, herbs, roots, and fruits.

Japanese calligraphy, in which lines and dots are remarkably important, represents one of the most popular arts in Japan. Calligraphic work is part of the learning of educated men who perform traditional tea ceremonies where the consumption of tea is promoted within a Zen environment. Inspired mainly by this Japanese art, the branding proposal utilized water and ink stains to represent the complexness and lightness of tea mixtures.

In the same way, vertical typographic arrangements based on traditional Japanese reading were employed, obtaining a balance between the compounding of classic and modern typographical styles.

The main emblem symbolizes the Japanese wind chimes "Fuurin," which is placed on doors and windows of homes in early summer. The color palette focuses on natural tones with red accenting details and gold foil rendering elegance and the importance of the care for each product.

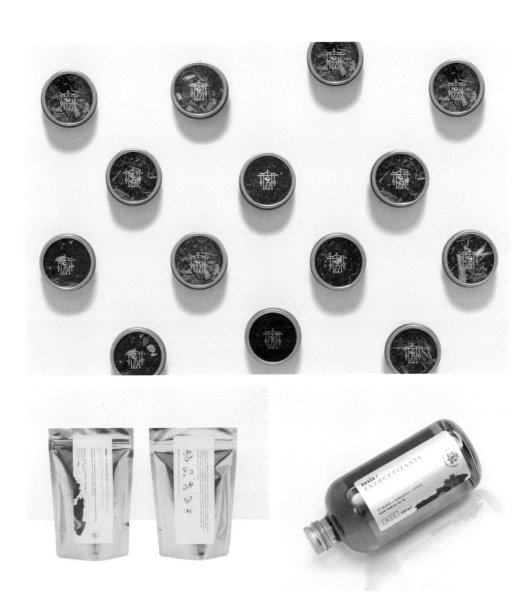

Lupchka

—

Client
Lupchka

Design
Mantik Branding & Creative
Digital Agency

Completion
2018

Minimalist design is about distributing visual information evenly, which brings value and places all the unnecessary detail on a second level. To achieve recognition in a world of minimalism, Mantik often brings sparks of surprises through color, texture, composition, and so forth.

At Mantik, the designers like to start new projects with strategy. In this particular case, the designers were assigned with naming the brand as well as position it on the market. Formed from combining "Lučka" (which means "popsicle" in Slovenian) and "Lupčka" (which means "kisses") the designers created "Lupchka"—a playful name that seemed to be obvious for a popsicle brand.

The designers opted to craft a distinctive, contemporary brand, with the white and colored labels reflecting the various flavors the packaging contains. The designers added glow-in-the-dark ink to distinguish the brand from the competitors' and to penetrate the adult demographic.

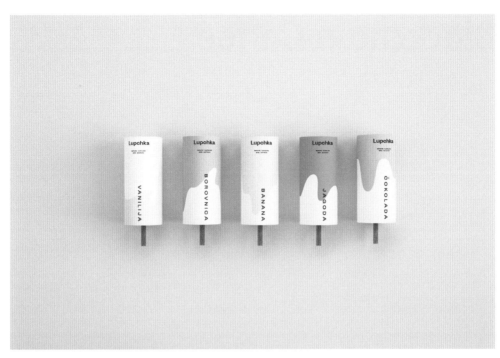

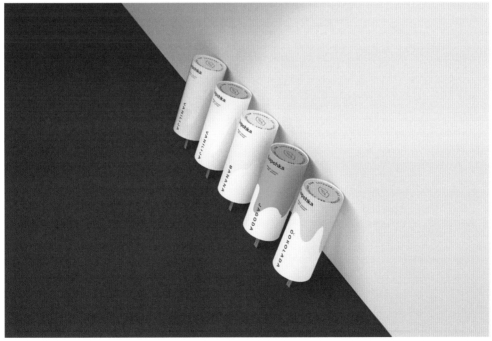

Blockchain Coffee

Client
Blockchain Coffee

Design
Stefan Andries

Completion
2018

To design packaging that stands out through a minimalist style, it's important to fuse both brand and packaging so that neither of them overpowers the other; rather, they complement each other.

An upcoming online coffee store, Blockchain Coffee combines the passion for coffee with the power of cryptocurrency. Because cryptocurrency and coffee are two totally different things, mixing them was a bit difficult initially, but when creating the login, the designer wanted to incorporate elements that were related to the blockchain (paths that interconnect to show the processes being made) and make these feel natural, not too techie, and more like an ancient style symbol or pattern.

The design of the packaging was rooted in the design and style of the brand, and contrast (cryptocurrency vs. coffee) was even more visible here. The designer wanted to solve this by keeping the design simple and using the pattern he created when composing the logo. The shape of the bag would be immediately recognizable as a coffee bag and the lines would interconnect and make it feel as if the blockchain links everything together. Because the front sides of the bags were white, the designer inverted the colors on the sides to offer a bit of contrast and also create a subtle continuation of the pattern.

PART 4

GEOMETRICS

Minimalist geometric forms are bold, sharp, and vibrant.

Geometric shapes complement and balance other elements and become a natural page layout grid or contrastive focal points. From physical boxes to visual patterns on labels, geometric shapes are the simplest and most popular graphic elements utilized in minimalist packaging design.

Raw

Client
Raw

Design
Mustafa Akülker

Completion
2018

The brand design for Raw—a detox drink company in New York—was inspired by fruits and vegetables from nature. Employing a crisp, minimalist aesthetic, designer Mustafa Akülker sought to convey the freshness of the brand through color choice to target the intended market of healthy drinkers wanting to detox.

The essence of minimalism is shown in this work. The designer said, "The simplicity I wanted to reflect, the effect I want to give, the brand expectation from me, and the modern air is all in this design." Printed on cool aluminum cans in yellow, green, pink, and blue with the bold, black "raw" logo, the products appear fresh and enticing, capturing the market demographic.

Hokkaido Powder Cube

Client
North Farm Stock

Design
Masayuki Terashima

Completion
2018

Offering cube-shaped cheese cookies of bite-sized proportions made with wheat from Hokkaido, Japan, North Farm Stock needed a standout packaging design to make its mark in the food industry and attract consumers.

Working with a small box with limited real estate for text and images, designer Masayuki Terashima decided to place the text information on the side of the box, with contrasting square patterns arranged on the sides and lid. Famous for snow in the Hokkaido region, an impression of glittering powdery snow was created via foil finishes on the squares on top of the box.

Aegean Shapes—small prints packaging

—

Client

The Round Button

Design

The Round Button
(Alexandra Papadimouli)

Completion

2016

For this project, the Round Button focused on combining minimal style and traditional elements of the Aegean Islands, then applying them to a variety of souvenir and decorative products. This specific packaging was created for the small prints of the "Aegean Shapes" souvenir series. The designer used a triangle since it is a shape found so often in the traditional Aegean architecture—the islanders use it in outside wall decoration, at their homes, and cobblestone walls.

A triangle was the ideal choice as a shape because it creates a steady box. The second step was to find a way to design the die cut so that the packaging wouldn't need glue or tape to close. It was very important that it would be possible to mail flat to the retailer and be assembled fast by anyone.

The designer chose 400gsm white and dark-gray textured paper, which proved to be a great choice for this project: the box stands beautifully. The colors harmonize with the general idea of the product: Aegean Islands equal to bright white light and dark-gray pebbles.

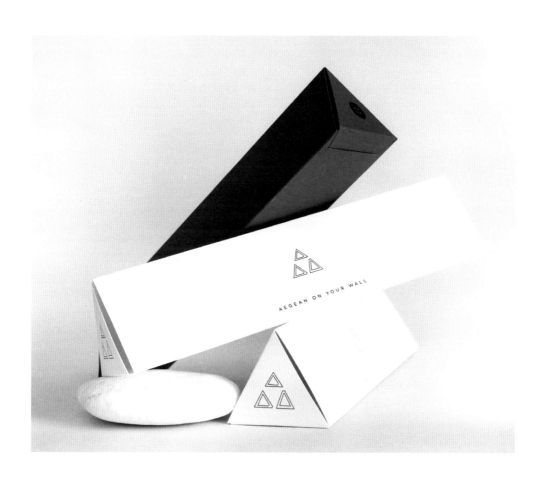

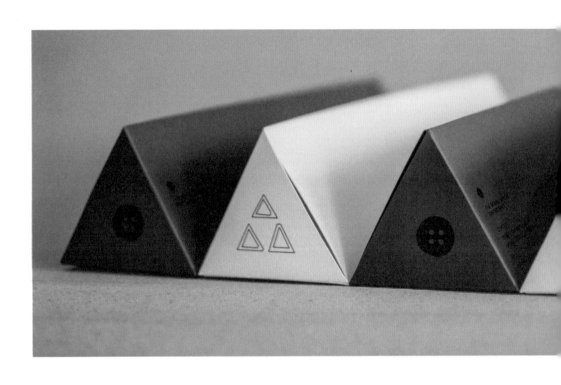

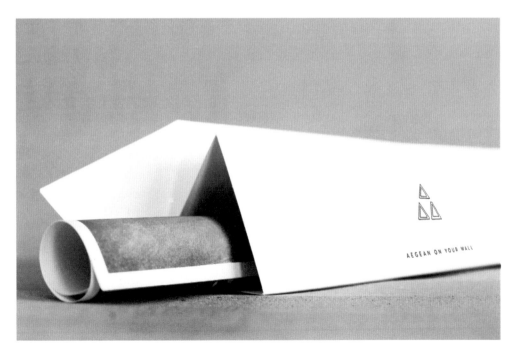

AEGEAN ON YOUR WALL

Rocky Mtn Chocolate packaging design

—

Client
Rocky Mtn Chocolate

Design
Wedge & Lever

Completion
2017

As part of Wedge & Lever's rebrand and strategic repositioning of Rocky Mtn Chocolate, Wedge & Lever completely revamped all of the packaging.

The two main categories of packaging were boxed chocolates and chocolate bars. The boxed chocolates focused on core minimalist principles to be a desired dinner party gift. The chocolate bars merged minimalist line work with vibrant colors to instill a youthful aesthetic that's more in line with the self-indulgent aspect of a chocolate bar.

It all came down to the details—perfectly placed and sized branding, just the right depth foil deboss, and materials that were soft to touch.

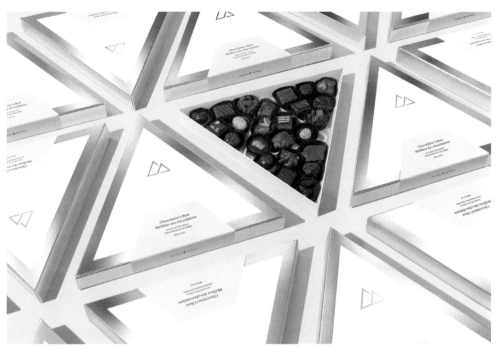

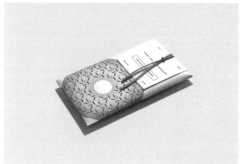

Buketo Wine packaging

—

Client
Cava Spiliadis

Design
Lazy Snail

Completion
2017

Buketo wine, named after the Greek word for bouquet, is created from grapes said to be picked like flowers and assembled for the perfect bunch every year.

Targeting youthful and open-minded wine seekers, the design brief called for a compelling, modern, and stylish label that would make the product serve as an impressive gift or new entry in the home cellar. Therefore, Lazy Snail played with this idea of a harmonious arrangement of elements and designed geometric compositions of fine lines vs. blocks of color and sharp edges vs. perfect circles. A larger composition is created if the two bottles—for the white and the red wine—are placed next to each other. The result is sophisticated and minimalistic, appealing to the target group, but also connoting the quality of the wines.

The Buketo logo is also geometric and minimalistic but builds on the idea of a bouquet in a more literal way. A bunch of three flowers was designed, referring to the aromatic character of the wines, as well as the three grape varieties used in making each of the wines in the series.

Old Style Vodka for Campus Warsaw

—

Client
Campus Warsaw

Design
Redkroft

Completion
2017

Campus Warsaw wanted a unique gift to hand out during special occasions and enlisted Redkroft to assist. At the time, Campus Warsaw, similar to other European branches, had no brand manual or any sort of solid visual guidelines. Each branch's design was based on intuition and a general sense of the brand rather than strict directives. This design had to fit in visually with a style that was still not fully established without straying too far away.

The designers decided that they couldn't ignore the location of the campus, an old—almost legendary—vodka distillery. As such, the idea of the gift being a bottle of vodka became obvious. The rest was a matter of aesthetics. In addition, it had to add some extra value. Silkscreen printed graphics and pleasant-to-touch papers were the weapons of choice. It would be more special if you could actually tell it was made with care and attention to details. This is hard to achieve on a factory line but the designers all agreed it should be made and assembled 100 percent by hand.

The colors chosen had a nice texture, somehow feeling warm to the touch. Silkscreen printed graphics added the pleasant embossment, which is felt with the fingertips. Manually applying the sealing wax was a bit painful, but it was just a small inconvenience in the whole process. To add some extra architectural detail, the designers decided to wrap the bottlenecks with a silkscreen covered calque, commonly used in architecture studios.

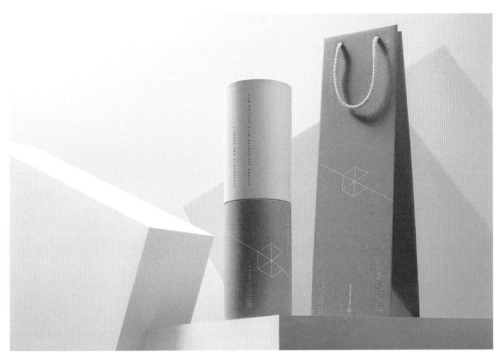

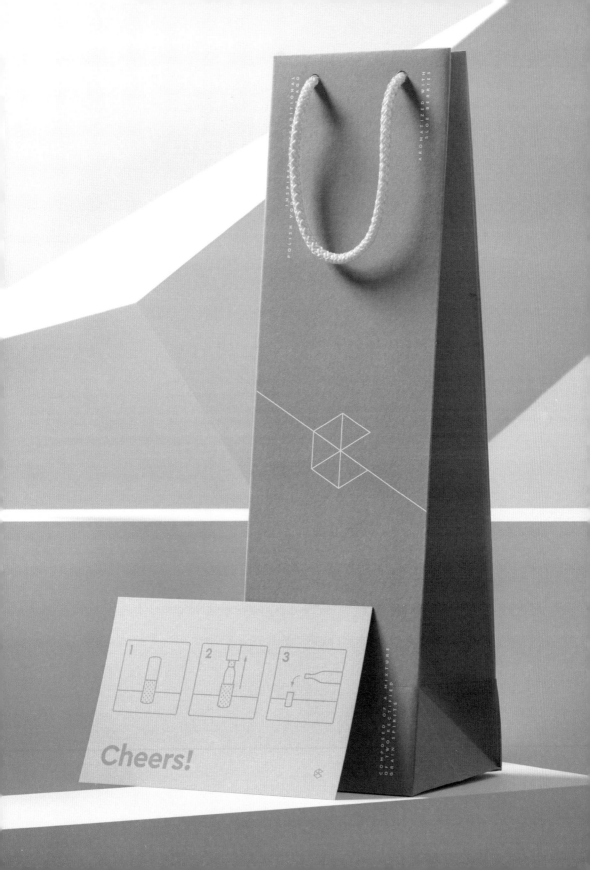

POLISH VODKA INSPIRED TRADITIONAL BRAND

AROMATIZED WITH SLOE BERRIES

COMPOSED OF A MIXTURE
OF TWO RECTIFIED
GRAIN SPIRITS

1 2 3

Cheers!

RYGR Brygghús brand identity and packaging

—

Client
Norske Bryggerier

Design
Frank Kommunikasjon

Completion
2018

RYGR Brygghús was a new local brewery by Norske Bryggerier and part of the company´s strategy to establish local breweries along the Norwegian coastline. Rygr is the original Norse name for the western region of Norway, Rogaland. Rygene is the old name of the inhabitants from this region, which suggests that they were growing rye. Rogaland is known for its rich Viking history, particularly the famous and great naval battle fought in Hafrsfjord around year 872, which led to the unification of Norway.

When looking for a distinct and recognizable design for the brewery and the different types of beer, designer Frank Kommunikasjon wanted to base the design concept on the proud Viking saga, their excellent craftsmanship, and their rich trading culture—sailing vast distances to faraway civilizations. Using shields for the main design, Kommunikasjon created a system of endless possibilities, embracing any new type of beer using both patterns and colors in combinations, which could be made into numerous and distinct series with a clear difference between the beer types.

The products were named after the regions from the Viking era. Hafr is Hafrsfjord, Jædr is the original name for the beautiful place Jæren, and Haugr is their name for the town Haugesund on the west coast. The logo is developed from a Viking axe to focus on the craftsmanship and the brewery is also located in Øksenevad in Kvernland Næringspark.

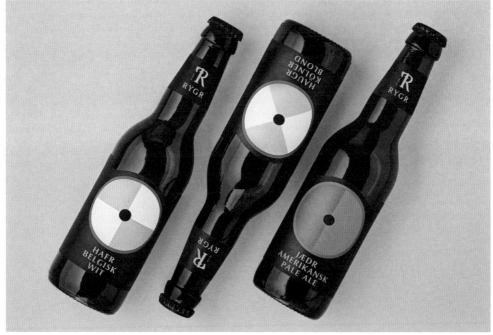

The Changer—Breaking Rules

—

Client

The Changer

Design

makebardo

Completion

2017

The name, The Changer, was selected as it not only literally describes what the product does—the very purpose of the syrup is to add it to other things to raise the flavor to another level—it also places the consumer in the position of protagonist to be a part of the product's history.

For the brand concept, the designers had two ideas: "transform to evolve" and "you break it, you own it." Why did the designers create such a disruptive concept? It was about enhancing the customer experience in a meaningful way to elevate the brand among its competitors with a unique personality.

For the visual identity, the designers embraced "essentialism," which isn't necessarily focused on reducing, but on what is most important. They selected a torn-paper style to follow the theme of "discovering new flavors by mixing," implying that by tearing the layers, consumers can discover new flavors. This was further emphasized with the use of copper foil.

The predominance of white with metal details on the label and box with the combination of a svelte bottle gives the product not only elegance but also graphic honesty to catch the eye of the consumer. The textured papers and touch of copper foil with an embossing finishing lend the brand to distinction, impact, and a crafted quality to the identity that reflects a personal approach to the brand.

Farac-Terzić Winery branding and packaging

—

Client
Farac-Terzić Winery

Design
Maji Studio

Completion
2018

Wine labels are usually all the same—with some illustrations and some gold letters. The packaging for Farac-Terzić shows how a simple twist with the label shape and typographic intervention can make a significant difference.

The visual identity and labels for Farac-Terzić wines were inspired by drywall structures that are very common in Dalmatia in Croatia, especially on the island of Korčula where the vine used in these wines is grown. Inspiration also came from the uniqueness of these wines because they are made of very rare vine sorts. And the outcomes are amazing, which can be seen from their placement in various competitions.

The design solution obviously had to be something different from the classical wine label and the result was simple. It consists of two things: a thicker paper with a distinct texture reminiscent of the karst landscape where this vine is grown, and the first letter of each wine's name, which is simulated by the forms of the labels themselves. This provides the system that allows expansion for the future wines that are yet to come.

○ ○ ○ ○ ○ ○ ○

PART 5

TRANSPARENT

You consume with your eyes first!

People love being able to see what they're buying. This not only requires competitive product contents but eye-catchy packaging design as well. The see-through area acts as negative space and the product's color is the background. Therefore, it is vital to minimize label design on the container to leave a noticeable larger area that allows consumers to focus on the product.

Without extra marketing, transparent minimalist packaging is more authentic and pure.

Zore Zalo packaging

—

Client
Zore Zalo P.C.

Design
busybuilding

Completion
2016

For the design of Zore Zalo—a new brand of Cretan tsikoudia, which is a type of pomace brandy similar to Italian grappa—the designers wanted the packaging to reflect the pure and minimal nature of its contents and highlight the premium character of the product.

The client's goal was to place a very traditional and local drink in premium bars around the world next to famous grappas and inspire mixologists to use it in cocktails just like any other pomace brandies. The name Zore Zalo means "difficult step" in the Cretan dialect, but also has the connotation of being dizzy, as in when someone is tipsy and staggers. Consequently, the name created was very local and traditional yet with an international ring to it. Through its strong, simple lines and understated appearance, the packaging successfully communicated the drink's premium character, helping it to stand out on the bar shelf next to well-known spirits.

For the container, the designers chose a completely transparent glass bottle with a simple black cap that is very pleasing to the touch. They illustrated the container with an exciting explosion of letters, corresponding to the explosion of flavor one experiences when tasting Zore Zalo. This could also be interpreted as a rush of emotions and an experience that cannot be described with words, and, thus, the words became visual elements.

Sweet Corn Soup

—

Client
Okamoto Farm

Design
Masayuki Terashima

Completion
2013

As the color of the product first shown by the client was thought to be very beautiful, designer Masayuki Terashima decided to make the product color itself the main visual element. Seeking a minimalist approach, Terashima thought it would be most effective to only use symbols to communicate to the consumer the product within, so he created a simplified vegetable icon and stuck on a transparent film seal. Offering a feeling of cleanliness and purity, the white icon gave the impression that the product does not contain unnecessary things.

Further to this, by attaching a black header to the top, the designer made the color of the product even more prominent.

Hokkaido Pickles

Client
North Farm Stock

Design
Masayuki Terashima

Completion
2017

As the pickles themselves were beautiful, designer Masayuki Terashima decided to make the packaging display the content as much as possible. However, since there were many similar products in the market, the designer wanted to create an icon that could differentiate this product from others. Using the image on yen, which looks like an eye, Terashima thought this to be an effective graphic. The icon was not printed on the bottles directly, but on seal paper of transparent film.

White Soy Sauce Ponzy

Client
Brasserie Cercles

Design
Masayuki Terashima

Completion
2017

Ponzu is a citrus seasoning used for shabu-shabu in Sapporo restaurant. Because the word "Pon" is interesting and easy to be remembered, the designer made it stand out as the main word. As ponzu is a seasoning for Japanese food, its packaging is usually made into a Japanese design style. However, the designer used a casual French brasserie style, designed with the alphabet to make it stand out. The text was printed on transparent film instead of the bottle directly because of the low-production quantity.

20/20 Riley

—

Client
Riley & Riley Ltd.

Design
Panos Tsakiris

Completion
2017

Riley & Riley sells a drink that has a mix of vitamins, rehydration salts, amino acids that break down toxins and caffeine, promises to prevent hangovers, and has one secret ingredient.

The main inspiration for this project was the effect that alcohol has on the brain and consequently the eyesight. The typeface and graphic elements were created based on this very idea. The name of the drink (20/20) was placed partially on the front and on the back of the bottle label. The more the user consumes the product, the more visible the logo gets. When the amount reaches the lowest point, the logo comes into full visibility. This subsequently means the user has absorbed all the required ingredients.

The most important thing anyone should do to avoid the next day's hangover is to rehydrate, which is the reason why the 20/20 logo was placed near the bottom rather than the middle or top of the bottle. The user is "forced" to drink to get the 20/20 effect and to fully rehydrate.

Additionally, the bottle is made out of recycled, recyclable thinner-than-usual plastic that would prevent any accidents from happening when someone is a bit tipsy.

The result was a system of clever graphic elements designed in order to support a bold, fresh, and elegant brand launch for the Riley brothers.

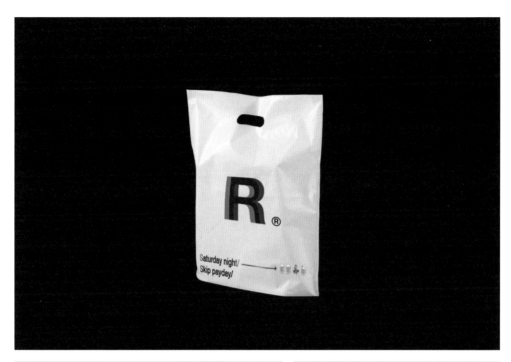

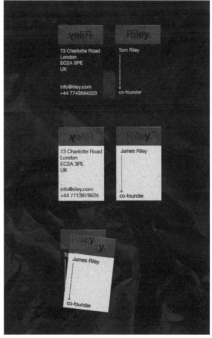

Maka

—

Client

Maka

Design

Anagrama

Completion

2017

Maka is a conscientious water-selling company with strong Mexican roots. As an eco-friendly and altruistic brand, they only use biodegradable materials for their bottles, and they support the local Nahua community in Mexico.

The objective of this project was to create a clear identity that reflected Maka's Mexican heritage and effectively transmitted their ecologic statement. So the designers created a clean and singular identity focused on distinguishing Maka as an all-Mexican brand.

The project started by taking Carlos Merida's artwork as inspiration. For the icon, the designers created an abstraction of a tzinitzcan, one of Mexico's most colorful and beautiful birds. The bottle design displays balance and transparency and works as a white canvas that allows different artwork to be included without altering its visual aesthetics.

○ ○ ● ● ● ○ ○

PART 6

SOPHISTICATE

Minimalist packaging is the ultimate sophisticate.

It maximizes elegant and unique visuals owing to less elements, less color, organic or simple geometric shapes and large negative space.

Office Stationery for Women

Client
Public Service Association
of New Zealand

Design
Regan Grafton (executive
creative director), Anne
Boothroyd (creative
director), Sam Henderson
(creative), Kent Briggs
(creative), Danny Carlsen
(design director), Luke
Harvey (photography),
Jamie Wright (retouching)

Completion
2018

The fact is that, on average, women in New Zealand earn 10 percent less than men for the same work. So, the designers came up with an equally ridiculous solution to deal with the gender pay gap—Office Stationery for Women, featuring a 13-hour clock and a diary with an extra month to help women work longer hours so they can earn the same as men.

Angry? Good. Because women shouldn't have to work extra to earn the same as a man. Office Stationery for Women calls for anyone who thinks equal pay is a basic human right to show their support for the PSA Worth 100% campaign.

Taking cues from the high street world of designer objects, the designers created (what appeared to be) a highly desirable stationery product. The packaging needed to be tantalizing and arouse curiosity so that people would want to open it and discover what was inside. A color palette of white, copper, and ironic "soft pink" was selected to evoke and reference the antiquated views towards femininity and the gender pay imbalance. It is a provocative gender reinforcing color to enhance stereotyping (of not just color but pay!).

13 MONTH DIARY
for women

WOMEN SHOULDN'T HAVE
TO WORK AN EXTRA MONTH
TO EARN THE SAME AS A MAN

If you agree that equal pay for women
is a basic right, then please share a pic
of this online and sign your name at
togethen.org.uk/worth100
#worth100

diary

OFFICE STATIONERY

for women

clock

OFFICE STATIONERY

for women

13 MONTH DIARY

for women

WOMEN SHOULDN'T HAVE
TO WORK AN EXTRA MONTH
TO EARN THE SAME AS A MAN

If you agree that equal pay for women
is a basic right, then please share a pic
of this online and sign your name at
together.org.nz/worth100

#worth100

diary

OFFICE STATIONERY

for women

clock

OFFICE STATIONERY

for women

Eskay

Client

Eskay Skincare

Design

Caterina Bianchini Studio

Completion

2017

Eskay is an all-natural, organic, and homemade skincare range based in Stockholm, originating from a vegan home. The most important thing for the company in the design was that clients could understand and read all of the ingredients used in the products.

For its branding, the logotype was created to have a delicate and considered layout with the inclusion of subtle design motifs. Wide-set characters created a more minimal and modern composition, creating a logotype that was ambiguous. Crafting a brand-new typeface for the logotype was essential to continue the fresh and handmade theme. The letters were drawn to have a slightly thicker width than height, creating a more modernist typeface.

The letters "S" and "K" were deliberately marked to reference the meaning of the word "Eskay," which is the phonetic sound of Saqera Kokayi's initials (Saqera is the founder of Eskay). It was essential to create a link between Saqera and the products, as the company began with the products being cooked up in her kitchen, ensuring every ingredient was hand-picked, locally sourced, and organic. This concept also ensured that all of the products had a personal touch.

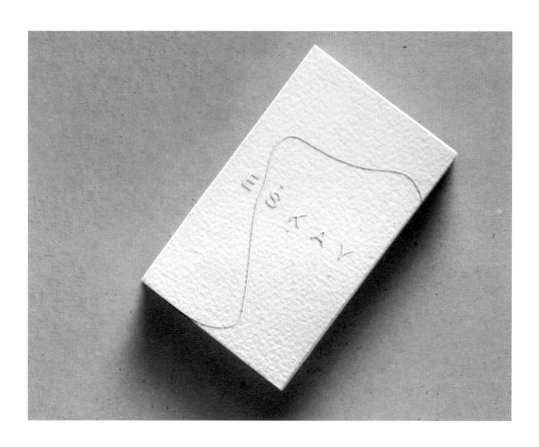

Verk visual identity and packaging

—

Client
Verk

Design
Studio Ahremark

Completion
2017

Verk is a Swedish watch manufacturer focusing on the essentials of timekeeping, inspired by the Scandinavian design heritage. Verk, which means "artwork" as well as "clockwork" in Swedish, has an artistic and minimalistic approach to the art of watchmaking, creating functional and aesthetic watches fitted for the modern man and woman.

Watchmaker Verk came to Studio Ahremark at the beginning of their journey as a company. They needed a strong visual identity and elegant packaging that would elevate them above the average watch brand and present customers with a luxurious experience. The resulting applications accommodated a simple yet bold visual system based on clean cuts, an all-gray color palette, and sharp typography in the form of Proxima Nova Alt.

Verk was very invested and helpful in the design process by giving clear briefs and explaining their intentions of the brand. Some directives and challenges were that the overall look and feel of the brand needed to be attached to the Scandinavian design heritage, which presently and historically has been heavily based around the idea of simplicity and functionalism. It's a long tradition of wanting to create design solutions that were easily accessible, democratic, and environmentally friendly.

Reduction helps in clarifying communicative aspects and as a bonus very often minimizes waste. The solution for Verk was also based on the idea

that the product should be the centerpiece and the packaging shouldn't be competing for attention; instead, they should blend together as if they were a part of the same object. In that sense, the packaging became the canvas and the watch became the painting. The reason behind the gold-foil stamping was that it would help the packaging catch one's eyes and by avoiding introducing any "artificial" colors to the mix and using colors that resemble materials already found on the watch, the designer's ambition was to create a design that complemented the watch instead of competing with it.

Good design is a result of understanding a brand and its audience. Minimalist is not a recipe for good design by itself, but reduction can be used to reduce a visual concept to its essence, and consequently find the thing that makes the right people notice the brand. Furthermore, the different printing techniques such as foil-stamping and the choice of materials were great ways to make it feel unique and attract attention.

Masyome

—

Client
J-Frontier Vestments Co.,
ltd.

Design
6Sense. Inc (Keiko
Akatsuka)

Completion
2018

In some instances, minimal packaging design—with a crisp and clean aura—can be more prominent and effective than bold packages. Upon commencing this project—for a brown-rice heat bag used to ease period pains—the designer envisaged a small box with a drawer, reminiscent of an underwear drawer, as the product is intimate and used only by women during a personal time of the month. Additionally, as the product is like a gift to the body, the designer crafted a ribbon into the design to make it appear somewhat like a present. The most difficult thing about developing a box is in selecting the packaging material. Since the item is a little bit heavy—a rice bag—it was important to make the box strong but beautiful; as such, an elegant, minimal design was necessary, with the slide-out function, ribbon, and rice paper insert making the product feel very gift-like and pampering, which is of huge appeal to the female demographic.

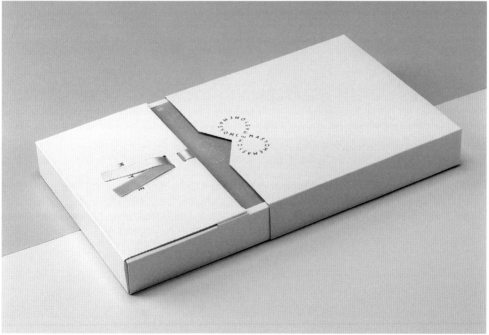

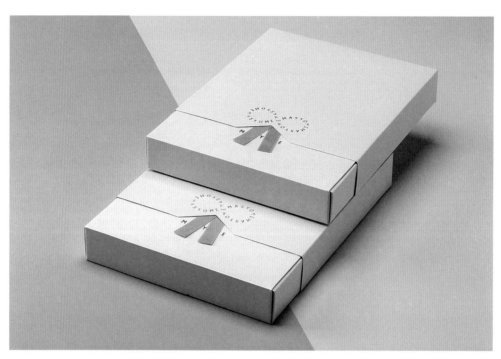

Gli Affinati—Scented Chocolate

—

Client
Sabadì

Design
Happycentro

Completion
2018

From the first chocolate cellar of its kind, Sabadì created a chocolate "scented" through a natural process. Thanks to its fat component, cocoa spontaneously absorbs the aromatic notes that surround it and maintains them over time. The chocolate is scented with seven elements for the production of unique chocolates in limited editions: tobacco, tea, flowers, herbs, spices, resins, and barrique casks.

The motto "less is more" perfectly embodies the sleek and elegant design Happycentro created for Gli Affinati, the latest revolutionary product by chocolate visionary Sabadì. Happycentro designed a series of cards inspired by botanic illustrations, a petite tin chest, and the simplest label: the copper hot foil typography describes the product, while in the dedicated blank space, the specific taste is in a handwritten style.

De Vry Distillery label and packaging design

—

Client
De Vry Distillery

Design
Whitespace Creative

Completion
2016

This distillery brand is centred on three core concepts: a shared passion for crafting, a love for country, and a quest for freedom of expression in respect of both taste and production. The initial client brief asked the creative team to interpret these facts into a concept that could be clearly and effectively communicated across all current and future brand contact points.

Minimalist packaging design has been a recurring trend, but at Whitespace Creative, design is not dictated by trend, but rather by primary design principles that are proven to be timeless—the most prominent being the principle of white space. However, this does not only refer to the literal blank space found within the design, but also pertains to what has intentionally been left out of the design. This results in designs that may look minimal, yet speak of intricate thought, elaborate strategy, and consideration as to the most important message.

The concept of this work was led by the fact that an Afrikaans distillery, located in the Free State, was producing a range of traditional English, Dutch, and Polish spirits. This fact was quite unusual and wonderfully ironic due to cultural differences between the origins of the spirits and the "boere" culture of the distillers. The three brothers were setting up a new perspective on these spirits that are in no way typically African. Their heritage was brought into the intentionally misspelt names that imply a humorous, tongue-in-cheek tone.

No7 Sauce ART

—

Client
Burger No7

Design
Caka Workhshop

Completion
2018

The designers were tasked with creating new packaging for a range of sauces for Burger No7, a growing food chain in Izmir, Turkey. Enlisted at short notice driven by the customers' constant pressure to innovate and the team's desire to freshen up the ordinary packaging, the client sought something that would clearly identify the product itself while also offering a modern style.

The result is a stylish range of packaged sauces in matt black, each with a splash-like feature indicating the particular sauce within by both color and font. From greens and yellow to convey mayonnaise and mustard to shades of reds depicting ketchup and hot sauce, each flavor is instantly identifiable, which spells success for both customers and staff in the fast-paced food industry.

BURGER
№7

GARLIC
MAYONNAISE

BURGER
№7

MAYONNAISE

BURGER
№7

RANCH

BURGER
№7

NO7 SAUCE

BURGER
№7

HONEY
MUSTARD

BURGER
№7

HOT SAUCE

BURGER
№7

KETCHUP

BURGER
№7

BARBEQUE

Surfing the sky

Client
Surfing the sky

Design
Shift

Completion
2016

Surfing the sky is an innovative consulting firm specialized in communication strategies and creators of the highly successful "Surfing the sky" methodology for long-term data retention.

A clean and modern graphic identity was developed as part of a wider brand update that included a refreshed version of the brands iconic logo bearing the firm's name and "surfer." A bold combination of white and blue colors with red accents conveyed the brand's relaxed, fresh, and forward-thinking attitude, while a secondary serif type and ordered layout nod to the deeply scientific and calculated analysis behind the firm's consulting approach.

The brand experience was further enhanced with every project's delivery in the form of a solutions kit incorporating the customer's company analysis, results, and data. This data was embedded within the kit in the form of a USB pill, alluding to Surf in the sky's scientific background and role as a healer or purveyor of results.

Ernesto Hermosillo
Sepúlveda

fundador
developer.

dirección.
Lázaro Cárdenas 1007,
Torre I85 #212
Residencial Santa Bárbara
San Pedro Garza García, N.L.

teléfono.
(81) 12 53 76 36

correo.
contacto@surfingthesky.com

surfingthesky.com

surfing
thesky

¿qué hacemos?

Una empresa dedicada a
consultoría innovadora en
comunicación estratégica.
Desarrollamos nuestra
metodología Surfing the Sky,
la cuál integra un sistema de
filtrado de información para
llegar al mensaje óptimo y
transmitirlo de la manera
más efectiva con retención a
largo plazo.

contenido.
usb, playera promocional
brochure.

talla.
s   m (l) xl

dirección.
Lázaro Cárdenas 1007,
Torre I85 #212
Residencial Santa Bárbara
San Pedro Garza García, N.L.

teléfono.
(81) 12 53 76 36

surfingthesky.com

The Line

—

Client
The Line (Marie Bottin)

Design
Fagerström

Completion
2017

The Line helps premium and luxury brands to develop visual projects through made-to-measure creative consulting. It also provides art buying and art brokerage services for individuals or companies.

In a world oversaturated by information, images, sounds, and so forth, it seems that with less elements you can get more attention than by shouting. The main difficulty or challenge was that the client knew a lot about design and art direction so they were very demanding in this sense. After a series of meetings with the client and an extensive phase of research, the designers got immersed in the world of art buyers and curators to learn about how brands live in this environment and how they communicate.

The inspiration for this brand was the relationship between art and the company, since art is present in most of the work developed by them. The visual identity sought to explore the use of lines in this discipline, playing with space and volume.

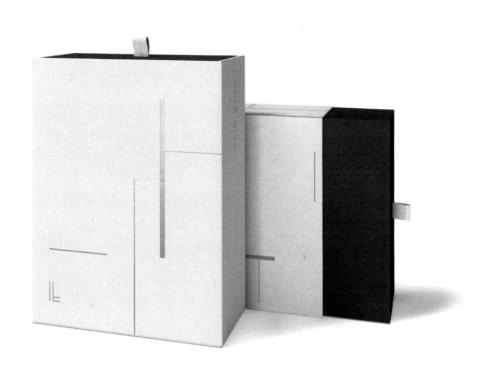

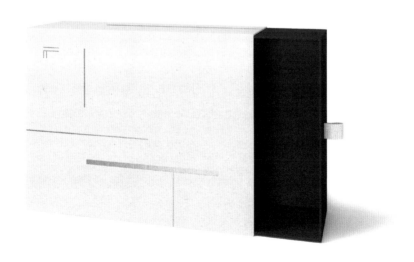

Gigo

—

Client

Sociedade Agrícola Casal
de Ventozela

Design

Gen Design Studio

Completion

2018

Portuguese winemaker Sociedade Agrícola Casal de Ventozela commissioned Gen Design Studio to create an identity and label for its new venture: a bold and modern wine brand grounded on the Douro legacy. Douro is a Portuguese wine region centered on the Douro River in the northeast part of Portugal.

From the beginning of the project, the designers knew it was important to relate the label to the name of the wine. Gigo is the name for the traditional baskets used in Douro for grape harvesting, involving carrying the grapes to the wine mill to be transformed into wine. To preserve this imaginary, the label was designed as an abstract representation of gigo (the basket) itself.

Exploring the relationship between form and content, it was intended to exalt the function of the basket, extending its usefulness to the bottle. The weave pattern execution served as the base for the label composition where the die-cut and transparent foil further enhanced the interweaving recognition and perceived quality of the product. The use of two different letters "g" (a double-story and single-story) offers a singular and memorable quality to the wordmark, and the choice of the wide Grotesk font gives it a bolder character.

Liqueur Black Milk package

Client

OK DEsign

Design

Kashkovskaya Oksana

Completion

2018

A unique liqueur packaging consists of two different color sides, which is the optical illusion created by two transparent glass bottles one inside of the other. The outer bottle is bigger than the inner one that has the decorative black pattern. The design challenge for the Liqueur Black Milk packaging was in creating a strong visual identity for the product based on the wholesome nature of the name and within the black-and-white color scheme.

Crafted from natural paper of medium thickness with a matte finish, the box design creates contrast and variety with all-black on one side displaying the white side of the bottle through a cut-out window, and all white on the flip side exhibiting the black side of the liqueur on the other. Consumers are invited to select from a choice of visuals, which could be reflective of their mood or the occasion. The two sides of the box are connected by the main graphic element of a cow's spots, which was designed in sync with the "milk" theme.

Also styled in black and white, the label juxtaposes the color palette with a white "B" on the word "Black" and is capped off by a wooden lock-cap bottle top with an organic resonance.

Enrejado

Client

Enrejado

Design

F33

Completion

2017

The project came to the designers with the name of the dry gin—Enregado—already registered as a brand by the client, which means "lattice" or "grille." The designers, as such, had to adapt and build around the preexisting name, which is more difficult than working from scratch when it comes to packaging as the concept has to be carefully considered to fit.

The designers played with the idea of a crow trapped behind wire because the name dragged the designers towards it inevitably. The raven seemed to have nocturnal connotations as wild as the gin itself and would be released upon opening the bottle and consuming its contents. Using the same colors for the bottle and the grating, the designers got them to merge, generating textures that do not obstruct the graphic itself.

The result is a gin that comes in a box with an ominous crow on the front, and on the side is a pattern of chicken wire. Inside, the bottles are also wrapped in a sleeve of chicken wire to enhance the experience for the consumer. The color palette chosen was simple—black and white only—which pops against the red used in their photography, while also giving it a classic, modern look that implies there is more than meets the eye.

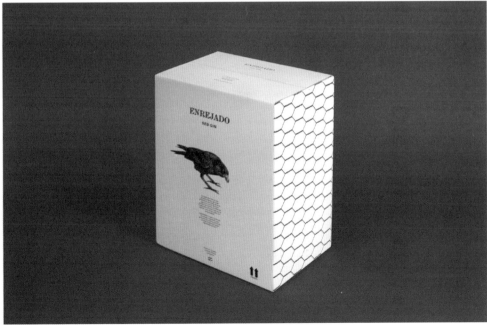

PART 7

BOLDER

Choose a single element (image, photo, logo, typography) and make it bold. Make it stand out. This standout element should give consumers a meaningful focal point. It could be the brand logo, product name, size, or type. The bold effect could be a larger size, more contrast, or the highest in the hierarchy.

GET RAW rebrand

—

Client
GET RAW AB

Design
SNASK (Matej Špánik)

Completion
2017

GET RAW is a health bar that is organic, vegan, gluten-free, and refined sugar-free. They contacted SNASK to do a full rebrand to modernize their visual identity and tone of voice.

The approach to the project was to not overcomplicate the packaging with unnecessary elements like photos of ingredients. The designers approached the packaging with the cheeky GET RAW philosophy, which is "junk-free snacking with no bullshit in the product and no bullshit in the design." Additionally, the design of this lifestyle brand had to work on social media with influencers and be likeable and shareable.

The designers sought to bring a bit of the raw feeling into the identity, which is why they chose a handwritten approach, taking some time to find the brush that would have the right feeling. The dynamic look goes well with an active and healthy lifestyle and also offers playfulness.

Hokkaido Poppy Corn Chocolate

Client
Sapporo Grand Hotel

Design
Masayuki Terashima

Completion
2016

As this product is corn wrapped in white chocolate, the designers adapted a literal interpretation in the packaging design. Upon a white box, a cob of corn has a layer of white partially overlapping it, which gives the illusion of white chocolate melting into the corn. Not only is this deliciously appetizing to consumers, once the see the word "chocolate" written prominently on the box and realize what the dripping white is depicting, it also offers an immediate visual explanation of the product inside.

The bold yellow of the corn stands out from the simplicity of the white box, while a clean font and modest rectangular box further promote the minimalistic success of this packaging design, letting the product within largely speak for itself.

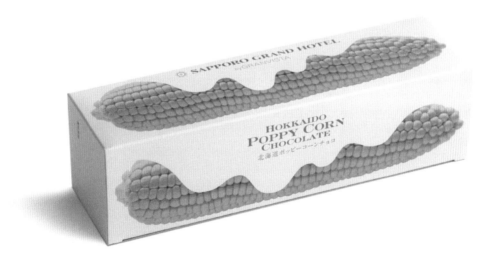

ΦΩΣ—Greek for Light

Client
La Petite Jumelle

Design
Semiotik

Completion
2017

The client's brief called for the design of a corporate gift to be given to their customers for the holidays. The proposal had to be original, tasteful, and in the Christmas or holiday spirit. At the same time, the final design was expected to take into consideration and highlight the client's business activity. The designers based their proposal on the concept of light, and on a relevant quote by renowned Greek author Nikos Kazantzakis: "The real meaning of enlightenment is to gaze with undimmed eyes on all darkness."

The designers created a scented candle in black. Unlit, it symbolizes the absence of light, while when lit, it "disperses the darkness." The candle itself, as well as the scent selected, both relate directly to Christmas symbolism. The candle was placed inside a specially made box, which was an opportunity for La Petite Jumelle to showcase their production capabilities.

The design objective was to make it stand out due to its symbolism and the messages it conveys, so simplicity in graphic and other design elements actually enhances its impact. It was also part of the brief to highlight the construction of the box, as it was manufactured by the client. The designers used materiality, such as foil printing, to add premium characteristics to a minimalist design, believing that reduction, when used effectively, can create just as much emphasis as addition.

Μήλο & Κανέλα

LaPetiteJumelle
.com

Production

SemiotikDesign
.com

Design

45h 15oz

JRINK Juicery branding and packaging

—

Client
JRINK Juicery

Design
Design Army

Completion
2016

Packaging's primary role is to serve the core thing it envelops. Successful execution of minimalist packaging, then, is about elevating the functionality and innovation of the thing inside. This often manifests itself as a celebration of negative space, monochrome color schemes, and restraint. Regardless of specific choices, the ultimate goal is to bring clarity and honesty to the design. In a world often inundated with gimmickry, this can capture the attention and imagination of the consumer. When done well, minimalist design can achieve an elusive sense of simplicity that feels both pure and timeless.

JRINK Juicery wanted to rebrand and find ways to cut costs on packaging. They were thinking about moving from glass to plastic bottles in an effort to save money and prepare for expansion. With so many new flavors coming to market, the question was how they would be able to reduce the cost of packaging, keep up with logistics, and create a simpler process while maintaining consistency for the brand.

With a focus on producing high-quality cold-pressed juices—made with 100 percent fresh, locally sourced produce—and providing a holistic cleansing

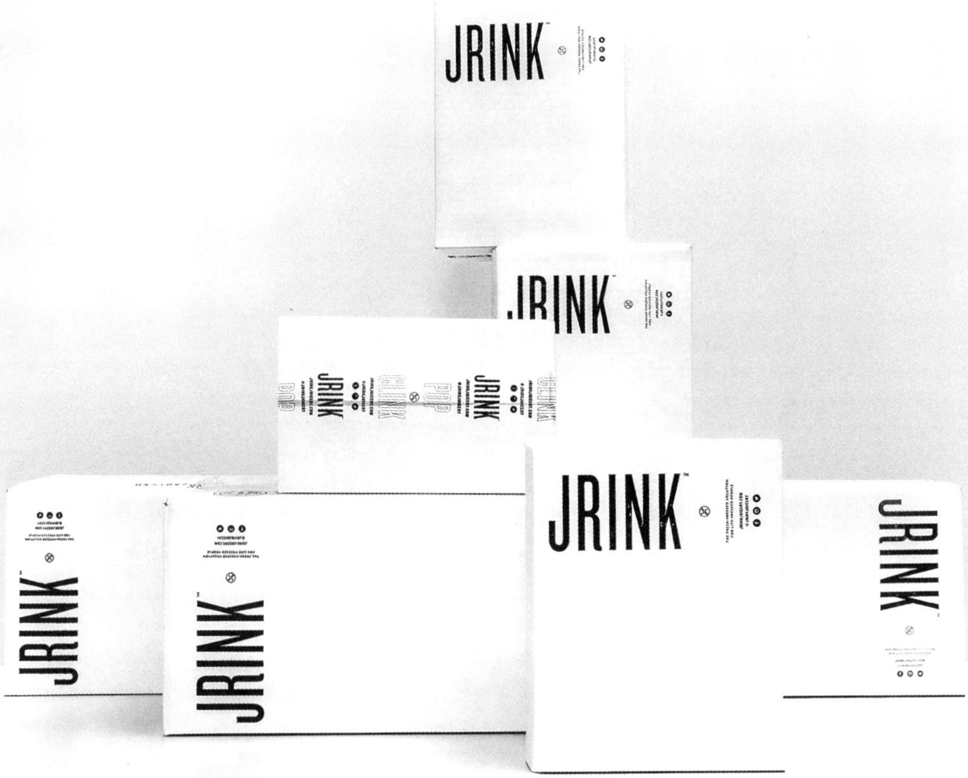
JRINK
JRINK
JRINK
JRINK
JRINK
JRINK
JRINK

experience, the designers believed that their "reboot" needed to reflect their healthy offerings and socially conscious outlook. They chose white for the packaging, de-cluttered the labels, and redesigned the logo to let the color of the juices speak to customers.

With the addition of their new delivery service, the designers also revised the boxes and packing tape to reflect the fun and playful personality of the brand, including lines like, "I have a jrinking problem" and "pop, clink, jrink."

Finally, the designers recommended they stick with recyclable glass bottles and moved the flavor information to the cap, reducing their need for 15 labels down to one and cutting productions costs. As such, every aspect of the client's business was filtered through a pure and authentic perspective befitting the brand's socially conscious lifestyle. Each new detail generated buzz on social media by showcasing their vibrantly colored juices and smart copy, growing the brand awareness and creating a more mature, clean look and feel.

BeFresh

—

Design
Erik Musin

Completion
2017

It's easy to make things more complicated, but to simplify a design is a difficult task. Working for an organic juice company, the designer sought to make the packaging of BeFresh as minimal as it could be, as the juice needed to stand out from other brands in the marketplace.

The simple solution was clear, square bottles, which allow the colorful juices to sell themselves, without the need to hide it under an overdesigned package. Customers can read all the information about the product through simple graphics and view icons on the back representing the mix of fruits or vegetables contained within, while the black cap on top of each bottle worked as a part of the bold logo.

Elsenwenger No.13

—

Client

Tischlerei Elsenwenger

Design

Studio Riebenbauer

Completion

2015

When it comes to a carpenter with a passion for their work, everything revolves around wood. So for the design of this brand identity, everything from business cards to brochures was crafted from wood. Using a bold font to convey information with a chalk-style graphic overlaying, the impression is that of a carpenter's handwriting, which further bolsters the identity.

To allow for all the wood's character to be experienced by clients, a unique element was developed—a series of "perfumes" in the form of bottles filled with various types and sizes of wood shavings to create an olfactory consulting tool. In this way, the client can choose the furniture not only by the look of the wood, but also by its smell.

Mathias Dahlgren

Client
Grand Hotel Stockholm,
Dafra & Mathias Dahlgren

Design
Essen International

Completion
2016

In a retail environment that is always busy and overwhelming, minimalist design tends to stand out by being clean and contrasting to the surroundings. The designers commenced the creative exploration of this project by getting to know Mathias. As one of the most famous and prominent chefs in Sweden, his cooking philosophy was a huge inspiration. It circuits around natural ingredients, simplicity, and joy.

The designers wanted to translate that into graphic design by using basic shapes, black-and-white colors, and focused typography. And being a design agency with strong Scandinavian roots, they were also inspired by Scandinavian minimalism and functionality.

The functionality is highlighted through the visualization of the product's frying surface in a simple, yet clear and distinct, circle. The functionality is further enhanced by highlighting the measurement of the products by placing them naturally and playfully, as one is used to seeing ingredients in a pan.

Milk

—

Design
Erik Musin

Completion
2017

Tasked with creating a new branding design from logo to packaging for milk, Russian designer Erik Musin from Saint Petersburg created something bold, minimalistic, and beautiful. The very first thing that came to his mind was a milk drop and the circle it leaves behind on a table. This was the initial inspiration from where he started to develop his idea.

Crisp and clean with a clear bottle displaying the white liquid inside and the simple "O" design paired with the word "milk," the striking simplicity of his design can't be overstated. In addition to this, Musin also sought to modernize the design of the bottle itself and, as such, crafted a twist-top design that eliminates the risk of dropping or losing the bottle top as it stays fixed and simply opens the hole through which to pour the milk, while keeping it fresh between usages.

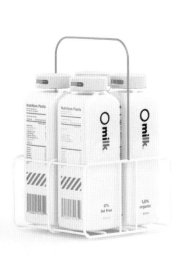

Alfredo Gonzales branding and packaging

—

Client
Alfredo Gonzales

Design
Anagrama

Completion
2016

Alfredo Gonzales is a Rotterdam-based brand focusing on offering unique stylized socks. The objective of this project was to update the brand identity, bringing it to a more modern context without sacrificing its personality. So the designers brought the brand's distinctive character to a more novel typographic language, as well as designing a brand-new handmade logo. Likewise, all illustrations were hand-drawn, underlining the unique style that distinguishes the band.

For the packaging, Anagrama designed a variety of simple yet timeless cases, focusing on the main product by fitting any style of socks. The client's indomitable spirit shines through this new tailor-made identity.

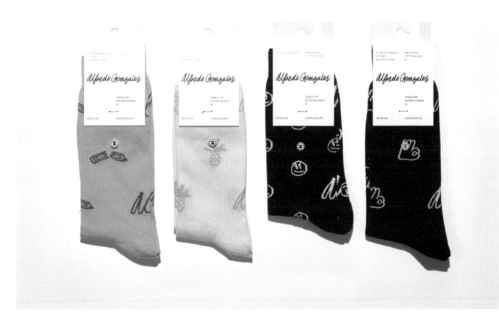

ALFREDO GONZALES
EST. 1983

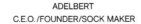

LIVE THE GOOD LIFE.

ADELBERT
C.E.O. /FOUNDER/SOCK MAKER

ADELBERT@ALFREDOGONZALES.COM
T. +31 452384028
GAFFELSTRAAT
ROTTERDAM, THE NETHERLANDS

ALFREDOGONZALES.COM

Index

Studio Ahremark
www.studioahremark.com
P 196

Studio Riebenbauer
riebenbauer.net
P 248

Tato Studio
tato.studio
P 046

The Round Button (Alexandra Papadimouli)
www.theroundbutton.com
P 146

Typical Organization
typical-organization.com
P 058

Veronika Levitskaya
made-studio.ru
P 124

Wedge & Lever
www.wedgeandlever.com
P 150

Whitespace Creative
www.wscd.co.za
P 208

Yuta Takahashi
www.yutatakahashi.jp
P 070

Published in Australia in 2019 by
The Images Publishing Group Pty Ltd
ABN 89 059 734 431

Offices

Melbourne
6 Bastow Place
Mulgrave, Victoria 3170
Australia
Tel: +61 3 9561 5544

New York
6 West 18th Street 4B
New York, NY 10011
United States
Tel: +1 212 645 1111

Shanghai
6F, Building C, 838 Guangji Road
Hongkou District, Shanghai 200434
China
Tel: +86 021 31260822

books@imagespublishing.com
www.imagespublishing.com

Copyright © The Images Publishing Group Pty Ltd 2019
The Images Publishing Group Reference Number: 1525

 A catalogue record for this
book is available from the
National Library of Australia

NATIONAL
LIBRARY
OF AUSTRALIA

Title:	Minimalist Packaging: Enhancing Creative Concepts
Author:	Chris Huang (introductory texts)
ISBN:	9781864708189

Printed by Everbest Printing Investment Limited, in Hong Kong/China

IMAGES has included on its website a page for special notices in relation to this and its other publications.
Please visit www.imagespublishing.com